Eat
Local,
Taste
Global

Eat Local, Taste Global

How Ethnocultural Food Reaches Our Tables

GLEN C. FILSON and
BAMIDELE ADEKUNLE

WILFRID LAURIER
UNIVERSITY PRESS

This book has been published with the help of a grant from the Canadian Federation
for the Humanities and Social Sciences, through the Awards to Scholarly Publications
Program, using funds provided by the Social Sciences and Humanities Research Council of
Canada. Wilfrid Laurier University Press acknowledges the support of the Canada Council
for the Arts for our publishing program. We acknowledge the financial support of the
Government of Canada through the Canada Book Fund for our publishing activities.
This work was supported by the Research Support Fund.

Library and Archives Canada Cataloguing in Publication

Eat local, taste global : how ethnocultural food reaches our tables /
Glen C. Filson and Bamidele Adekunle.

Includes bibliographical references and index.
Issued in print and electronic formats.
ISBN 978-1-77112-313-6 (softcover).—ISBN 978-1-77112-314-3 (PDF).—
ISBN 978-1-77112-315-0 (EPUB)

1. Ethnic food industry—Ontario—Toronto. 2. Vegetables—Ontario—Toronto.
3. Local foods—Ontario—Toronto. 4. Food security—Ontario—Toronto. I. Filson, Glen C.,
1947–. Political economy of ethnocultural vegetables in Canada. II. Adekunle, Bamidele.
Greater Toronto Area preferences for ethnocultural vegetables.

HD9333.C33O5 2017   338.1'9713541   C2017-902155-9
                                                   C2017-902156-7

Cover design and cover image by hwtstudio.com.
Text design by Janette Thompson (Jansom).

This book is printed on FSC® certified paper and is certified Ecologo. It contains
post-consumer fibre, is processed chlorine free, and is manufactured using biogas energy.

Printed in Canada

Contents

List of Tables

List of Figures

Acknowledgements

Work on this book began in May 2009, when we received funding to do a survey to determine the demand for ethnocultural vegetables in the Greater Toronto Area (GTA). Most of this funding came from Ontario Market Investment Fund of the Ontario Ministry of Agriculture, Food and Rural Affairs (OMAFRA) in addition to support from FarmStart, the Greater Toronto Area Agriculture Action Committee (GTAAAC), and the Toronto Food Policy Council (TFPC). To carry out that survey we employed ten graduate students, who along with Glen Filson, Bamidele Adekunle, and Sridharan Sethuratnam, formed part of a research group we called ECVOntario. Two years later ECVOntario teamed up with Vineland Research and Innovation Centre and FarmStart to receive additional OMAFRA funding to determine the value, market size, and evolution of ethnocultural vegetables. We secured more OMAFRA funding from the Agri-food Rural Link Knowledge Translation and Transfer program to increase vegetable producers and retailers' awareness of the demand for ethnocultural vegetables (ECV) and promote Ontario production of ECV.

ECVOntario's research could not have been carried out without the diligent efforts of Joy Sammy, Helena Kifle, Asumani Serugendo, Pradip Dey, Nichele Palen, Jeremiah Saringe, Dingfei Li, Andrew Filson, Shawn Filson, Morgan Sage, Kur Mayen, Meghan Bloom, Keteh Amba, Steve Gitu, Keisha Davis, and Joel Aitken.

The primary authors of this book include Glen C. Filson, Ph.D., retired professor, now adjunct professor, University of Guelph; and Bamidele Adekunle, Ph.D., adjunct professor and Special Graduate Faculty, SEDRD, University of Guelph, Guelph, and Contract Faculty, Ted Rogers School of Management and associate member, Yeates Graduate School, Ryerson University, Toronto. Contributors to individual chapters include Sridharan Sethuratnam, Ph.D. student, Geography, University of Guelph, and director, California Farm Academy, Center for Land Based Learning, 5265 Putah Creek Road, Winters, California; Dario Cidro, Ph.D., consultant to 3rd International Conference on Integrative Disaster Risk Reduction Management, Eastern Samar State University, Brongan, Eastern Samar, Republic of the Philippines; Yasantha Nawaratne, M.Sc., assistant manager (Operations) at Dominion Citrus, Meschinio Banana Company (Division of Dominion Citrus); Christine Kajumba, M.Sc., nursing consultant, Ottawa; Monika Korzun, Rural Studies Ph.D. student, University of Guelph; and Frances Dietrich-O'Connor, M.Sc., human environment consultant at Shared Value Solutions Ltd., Guelph, Ontario. Morgan Sage produced the index and helped throughout the research process. Special thanks are also due to copy editor Valerie Ahwee.

Three anonymous reviewers provided excellent feedback, as has Wilfrid Laurier University Press's senior editor, Siobhan McMenemy.

Introduction

Glen C. Filson

This book identifies the demand for culturally appropriate food and assesses the role of vegetable value chains in the latest international food regime. The latest international food regime is explained in detail in Chapter 1. It refers to the development of a newly dominant set of international food chains controlled by large multinational corporations from producers to assemblers, wholesalers, and retailers that connect food trade between and among Northern and Southern continents. It ascertains the growth potential of more local production of ethnocultural vegetables and additional market segments for local farmers. It also assesses the health potential of these vegetables, especially if and when they can be produced locally. As well, it examines alternative forms of local production and consumption, including farmers' markets and community shared agriculture, to assess the prevalence of non-white immigrants and their degree of provision of ethnocultural vegetables. We then suggest ways to shorten the long-distance vegetable value chains. In order to achieve this shortening we employ a political, economic, and class analytic framework identifying contradictions, including class conflicts that are changing access to these vegetables. Some of this will require policy changes (discussed in the final chapter) to enhance people's access to a greater variety of healthy food.

The fourth largest metropolitan area in North America, the GTA has seen its demographics transform dramatically. Toronto is now the world's most multicultural city. Not surprisingly, there has been a spectacular proliferation of ethnocultural food available in restaurants, ethnic stores, supermarkets, and farmers' markets, though formerly there was little reliable data available about the demand for specific ethnic vegetables.

More horticultural producers are recognizing the opportunity to produce for these market segments as more is known about which of these vegetables can be grown in Canada. Production of ethnocultural vegetables is now growing in backyards and community gardens, the Holland Marsh, Niagara, and north of Lake Erie, among other places, but more can be done to meet this demand locally, so we provide a framework for determining what forms of improved agricultural extension efforts, use of social media, and policy changes can enhance the provision of these healthy, increasingly culturally appropriate food.

Our initial collaborators, the Toronto Food Policy Council (TFPC) and the Greater Toronto Area Agricultural Action Committee, have played an important role in helping to connect immigrants with their preferred fresh food, in part by fostering interest among local farmers in producing ethnocultural vegetables. The TFPC, for instance, created the Toronto Food Charter of 2001, which advocates "the right to healthy, affordable, and culturally appropriate food" (quoted by Friedmann 2007, 390).

Our collaborators asked us to determine which vegetables were most in demand and document the cost. We began our research by focusing on the demand for ethnocultural vegetables in the GTA. This developed into an interest in how many of these ethnocultural vegetables could be grown locally given that most of them are currently imported.

We realized that understanding the main ethnocultural vegetable supply chains required us to identify both the international food regime that dominates ethnocultural vegetable procurement, as well as the local food movement, which is playing an increasing role in the production process. This includes the vertical and horizontal linkages involved, including the social and technical relations of production and distribution. As will be seen in Chapters 1 and 3, the supply chains for ethnocultural vegetables consumed in Ontario still originate primarily via the corporate food regime, that is, the international food chains that are dominated by multinational corporations. However, they are increasingly grown organically in Ontario even though to date their production remains relatively small scale.

Because of the importance of the political and economic elements of ethnocultural vegetable value chains, we employed a political economic analysis of how ethnicity and its interaction with social class affect people's access to their culturally appropriate food (Chapter 1). This involved analyzing the relationship between the owners of ethnically appropriate vegetable production, distribution, and marketing and those who manage and work within various components of the value chains. Access to culturally appropriate vegetables is also a function of the type of grocery, mainstream or alternative, where people shop for food. We assess these issues as well as the health benefits of eating these vegetables. Sadly, recent research has shown that most of Toronto's newcomers arrive in a relatively healthy state only to experience a decline in health some time later (*The Global City* 2011). Lack of access to their culturally preferred fruits and vegetables is a major reason for this.

Because the context of our research is economic, sociopolitical, sociological, and anthropological, the political economic approach was perceived to be broad enough to grasp the relation of ethnocultural vegetables to emergent food regimes, yet specific enough to help us understand how the changing class structures and ethnic and demographic processes bring about these dramatic shifts in consumption. This methodology is elaborated in Chapter 1 prior to explaining why, for instance, we preferred to use the term "ethnocultural vegetables" instead of the alternative term "world crops," which is used by some other researchers also studying these phenomena. This required that we differentiate national (English, French, and First Nations) from ethnic groups within Canadian history and ponder the appropriateness of increasingly popular terms like "interculturalism" and "transnationalism." In short, Chapter 1 explains the underlying political economic framework of our analysis, which uses historical class analysis of international ethnocultural food value chains that affect people's access to food sovereignty. Food sovereignty exists when culturally appropriate food is available to ethnic groups within their new home community.

Chapter 2 examines the implications of the demand for ethnocultural vegetables by South Asian, Chinese, and Afro-Caribbean Canadians in the Greater Toronto Area (GTA) and puts this in historical perspective. Data collection for the demand survey reported on in Chapter 2 took place in 2009 and 2010. Since then the information collected has been used to write three journal articles on each of the three largest ethnic groups in the GTA that together compose 46 percent of the overall population (2011) (Statistics Canada, 2013).[1] After we determined their total demand for ethnocultural vegetables as well as which vegetables each group preferred, we worked with southwestern Ontario horticultural researchers to identify which of these popular vegetables—such as okra, bok choy, Asian eggplant, and smooth amaranth—can be grown locally to increase people's access to them.

The GTA is home to more than six million people, including large multi-ethnic populations within Mississauga and Brampton. The rising demand for locally grown food has created a need to better understand how ethnocultural vegetables consumed by immigrants in their respective countries can be incorporated into our Canadian diet. Growing them locally would create new market opportunities for our producers. We investigated the market for these vegetables in the GTA and their short- and long-distance value chains.

Distinguishing differences among food regimes and understanding ethnocultural vegetable value chains (Chapter 3) enable the identification of their scope and context, the opportunities they present for healthier, culturally diverse Canadian diets, and the market sectors they make available to small-scale organic farmers as well as to commercial vegetable producers. We identify the main contradictions within international and local ethnocultural vegetable value chains. For instance, Ontario's commercial growers are typically of European descent so often they don't know how to grow most ethnocultural vegetables. Even if they do, they are usually not sure how to package and market them. On the other hand, it is difficult for small growers to sell their ethnocultural vegetables at the Ontario Food Terminal (OFT), except at the OFT Farmers' Market, since they lack sufficient volume and quality control to obtain a contract with appropriate wholesalers like Ippolitos. These mainstream wholesale and retail markers and the smaller, short-distance value chains that provide fresher, local produce mostly at farmers' markets, direct farm sales, and community shared agricultural operation are identified in Chapter 3.

By analyzing the relative pricing of ethnocultural vegetables at different types of wholesalers, ethnic stores, and supermarkets we attempt to ascertain the impact of globalization, immigration, and the retail market structure on the consumption of ethnocultural vegetables. Chapter 4 considers the importance of culturally appropriate vegetables to people's culture, faiths, and socialization. People's accessibility to this produce has increased with globalization, immigration, and evolving food regimes. We also reflect on the retail market structure where products of long-distance value chains and short-distance chains compete for shelf space with a range of different displays, qualities of packaging, and pricing.

Chapter 5 assesses farmers' markets and their degree of inclusivity. It addresses their advantages, such as freshness and consumer embeddedness, while asking why, for the most part, ethnocultural vegetables are generally not available in these markets. However, ethnic stores, restaurants, and even mainstream supermarkets are increasingly providing ethnically diverse food. Chapter 5 concludes with a review of the extent to which ethnocultural vegetables are crossing over to traditional Canadian ethnic and national groups.

The impressive growth of community shared agriculture (more than 250 now operate in Ontario) parallels the growth of farmers' markets. Their expansion is no doubt connected to opposition to the corporate food regime and the "industrial diet" (Winson 2013). Ignatius Community Shared Agriculture, for instance, often has ethnocultural vegetables. In some cases ethnocultural vegetables can be obtained via community shared agriculture but, as we show in Chapter 6, this access remains limited.

In acknowledging the rising interest in growing the specialty crops for these ethnic groups, Elford (2013) cautions that our (Adekunle et al. 2010, 2011, and 2012) study "was for all vegetables noted within the study groups, not just specialty vegetables. If one does not clearly recognise this point and review the crop list in the study, the potential market for specialty vegetables may be wrongly interpreted and possibly over-inflated." This cautionary note notwithstanding, demand for these vegetables has grown considerably in the past seven years since the survey was conducted.

The final chapter of this book reviews the political economy of ethnocultural vegetable production, distribution, and consumption, recognizing both the dynamism and strength of the corporate food regime, which has swept aside country boundaries with the rise of free trade and international sourcing of vegetables, on the one hand, and locally produced, in-season food, and import substitution production on the other hand. It identifies the most important contradictions within these vegetable value chains and suggests some remedies.

The international corporate food regime is dominated by fast-consolidating agro-transnational corporations (Weis 2007) yet many alternative food, social justice, and food sovereignty movements, including organic, local food and community food security movements, have also recently expanded. Each of these systems of procurement has distinct value chains, though inter-mediate chains exist as well. How the contradictions between these food systems evolve and transform both what and how we eat also determine whether we can eat local yet still taste global.

CHAPTER 1

The Political Economy of Ethnocultural Vegetables in Canada

Glen C. Filson

Ethnocultural Vegetables

Ethnocultural foods are similar, but not exactly the same as what horticulturalist Brownbridge (2017) calls "world crops" and others like Tilman et al. (2011) call "global food." World crops and global food refer to all the world's food, whereas the term "ethnocultural vegetables" as used here denotes only those vegetables that ethnocultural immigrants to the Greater Toronto Area perceive to be indigenous to their cultures. Considering that of all the food that people eat "worldwide, 69 percent is foreign . . . everybody's connected to everywhere else to some degree or another" (Harvey 2016). Consequently, there is necessarily some overlap among all vegetables. Maize was originally bred and developed by Mexicans, then exported globally so that today, for instance, maize is considered Zambia's national food staple (Thurlow et al. 2008). It is therefore an *ethnocultural vegetable* for our Afro-Caribbean respondents (Chapter 2), though it has been a *national* vegetable in Canada for thousands of years as part of indigenous North America's "three sisters"—corn, beans, and squash. When Italians first immigrated to Canada, they brought cilantro, broccoli, rapini, artichoke, and eggplant. Most of these vegetables are considered mainstream in Canada because they first made an appearance in New France, Lower or Upper Canada. Yet there is overlap with other vegetables as artichokes come from Latin America and slightly different eggplant species are African (round, white "garden egg") or Asian (long, thin, "purple eggplant"). So we refer to the latter two as ethnocultural vegetables even though they overlap with European vegetables, which arrived in Canada many years earlier and are now considered mainstream.

Canada relies most on "foreign" food (Harvey 2016). Even though the production of ethnocultural vegetables continues to be relatively marginal in contrast with most local farmers' production for the world market, the local production of these vegetables continues to grow despite the difficulties facing commercial producers, such as their lack of access to minor crop pesticides, the consequence of which is that most ethnocultural vegetables must be grown organically in Canada. Because the market segment for ethnocultural vegetables is growing along with immigration, their importation via the latest corporate food regime has, on the other hand, continued to grow along with Canadian consumers' demand for them.

The term "ethnocultural vegetable" is also based on distinctions among the meanings of ethnicity, culture, nation, and social class. Ethnic identity has been defined as "a common cultural tradition and a sense of identity as a traditionally distinct subgroup within a larger society; and possibly possession of their own language, religion and distinct customs" (Anderson and Frideres 1981, 37). Individual immigrants from particular ethnic groups also vary in the degree to which they adapt to the host culture and oscillate back toward "situational ethnicity" in a different context (Muller et al. 2009). The importance of ethnicity and culture to people's identity is one of the reasons why it is imperative to preserve indigenous and European Canadian food cultures as well as ethnic food cultures. The United Nations considers access to people's culturally appropriate food to be a basic human right.

Acculturation is the process by which someone of one ethnic group adopts the culture of another ethnic group. This is a mutual process to some degree, though it varies from country to country and region to region. Some countries consciously try to assimilate new ethnic groups into the dominant "melting pot" culture, while other countries, including Canada, do not. Pierre Trudeau's Liberal government policy (1971) of "multiculturalism within a bilingual framework" marked a change in the federal government's practice of attempting to assimilate people of different ethnicities and indigenous nations into the dominant British and French cultures. Of course, Canada still has a long way to go before it can be considered truly inclusive and tolerant.

Such ethnic characteristics as language, culture, folklore, ethnic self-consciousness, and a sense of common origin maintain considerable stability, not just across social classes, but from one mode of production to another, such as from feudalism to capitalism (Bromley 1974; Hall 1986; Smith 2000).[1] While the term *ethnicity* is derived from the ancient Greek word *ethnos*, which referred to a variety of forms, including peoples, nations, and even races, the modern term *ethnicity* was first used by David Reisman in 1953 (Filson 1983; Glazer and Moynihan 1975).

Hamlett et al. (2008) point to the importance of always contextualizing ethnicity so that it is not divorced from people's social class, age, and gender—so that ethnicity is not considered merely in and of itself. In using the term *ethnicity*, Hamlett et al. (2008, 99) caution that "ethnic identity, like class, needs to be seen in the context of other identities and power structures such as gender." For instance, "if [South Asian] women worked outside the home, they were more likely to invest in convenience food, increasingly available in supermarkets and self-service stores" (Hamlett et al. 2008, 110–11). Whether or not women work outside the home is partly a function of religious and cultural traditions as well as local employment opportunities, their educational level, and age, and these factors influence whether men or women are the primary food shoppers.

Dietary preferences usually have more to do with ethnic than social class differences and similarities as ethnic groups exist in all social classes (Laclau 1977; Wright 1985). The prevalence of intra-ethnic conflict and the variability of ethnic relations and descent rules used to assign ethnicity to people of mixed ancestry provide a basis for rejecting the idea that ethnicity is biologically based. Unlike race, ethnicity is acquired through one's socialization (Bonacich 1980).

Culture, a feature of both ethnic groups and nations, refers to "the anthropological and extended sociological use to indicate the 'whole way of life' of a distinct people or other social group" (Williams 1981, 11), including the social practices they engage in that influence the unconscious traits of their subjectivity—their character structures or personalities. In this sense, culture is distinct from the conscious beliefs, ideas, and attitudes of a people. To be part of their consciousness, the ideas and attitudes of a people are those aspects of which they are aware, whereas people's culture has a deeper impact on their subjectivity.

When one or more ethnic groups form a historical community of social development that has a common form of economic development combined with common cultural features, language(s), and patterns of thinking about their environment, these people constitute a *nation*. To exist, a nation must have a common territory, common languages, common customs, and common psychology. Nations are often classified by their level of economic development, such as socialist nations, advanced capitalist nations, imperialized and underdeveloped nations, all of which subsume particular social and technical relations of production, social classes, and strata (Bromley 1974).

Nations are therefore similar to *ethnic groups*, except that nations also comprise territory, something that may have been lost by particular ethnic groups at a specific level of development. Nations also possess governments. Of course, the uneven development of capitalism, combined with the fact that several modes of production are often articulated together in a specific nation's social formation (specific economic structure), means that particular nations may be comprised of various levels of economic development, which is clearly true of Canada. Thus a nation comprises the ethnic splinters of other nations, the national groups and social classes that exist in a particular territory (Fedoseyev and Schneierson 1977).

The first Europeans to interact with North America's Aboriginal peoples were confronted by already existing nations such as the Iroquois Confederacy. In turn, the Iroquois Confederacy was comprised of distinct communities like the Seneca and Mohawk, each speaking their own dialects and occupying particular territories. The First Nations, French Canadians, and English Canadians each have a legitimate claim to territorial, linguistic, and cultural rights, which are recognized in the Canadian Constitution and Charter of Rights. Kymlicka's (1995, 79) theory of multicultural citizenship also identifies the English, French, and Aboriginal peoples as national minorities in contrast with ethnic minorities "who have left their national community to enter another society." Ethnic minorities have come voluntarily and are spread throughout the nation. As will be seen later in this chapter, this conception raises important challenges for the concept of *transnationalism*.

As Haque (2012, 244) summarizes Kymlicka's position:

> The key distinction is that national minorities, at the time of their incorporation, "constituted an ongoing societal culture and may have had rights regarding language and land use" (Kymlicka,

1995: 79), whereas ethnic minorities came "voluntarily," are not "nations," and do not occupy homelands but are scattered throughout the nation-space.

And in Canada, these ethnic minorities, unless elderly, must learn English or French. Despite attempts to force First Nations to speak only English, French, or both, at least within their own Canadian territories, they should be free to speak their native languages, but in most cases are not.

When immigrants to the United States (and probably Canada) become acculturated to the dominant cultural norms, they may adopt many unhealthy behaviours such as over-consumption of processed food containing salt, sugar, and fat (Neff et al. 2009). Thus acculturation in this sense is not a good thing, though it can also have healthy, *transcultural* or *cross-cultural* crossover effects such as when Canadians of European origin begin to eat bok choy, okra, bitter melon, and other ethnocultural vegetables that, when fresh, have considerable nutritional value. As Hamlett et al. (2008, 109) put it, researchers have increasingly found that acculturation has become "a useful way of understanding 'cross-cultural consumption.'"

Beyond freshness and the need to retain vegetables' health-promoting, nutritional qualities, there are many other advantages to producing this food locally. Such benefits include reducing our carbon footprint, generating additional income for local farmers, and providing the conditions that will enable Canada to be a truly multicultural society while supporting healthy lifestyles for Canadians.

There is somewhat of a political economic tug-of-war between those controlling most of what we eat and those wanting a more sustainable agriculture and local access to fresh, healthy food. This struggle occurs along the value chain from producers to processors, and distributors, wholesalers, and retailers all the way to the consumers. We focus on ethnocultural vegetable provision in part because there is growing demand for this produce from our increasingly multicultural society as newcomers seek fresh, culturally appropriate food. There are also many contradictions within the value chains between the large international merchants and producers on the one hand and the plethora of smaller, usually ethnic buyers and sellers as well as between and among the various classes of people within and across ethnic groups from the North and the South.

Whether or not women are the primary food purchasers—and this varies from culture to culture—the connection between consumption patterns and ethnicity in Canada is not well understood (Abdel-Ghany and Sharpe 1997), but increasing evidence of the connection between class and consumption is becoming known. Poorer people, for instance, tend to consume cheaper processed food and fewer more expensive fruits and vegetables (Winson 2013; Neff et al. 2009). Wealthier people can afford higher priced fresh, organic produce. As Darmon and Drewnowski (2008, 1007) observe, "Whereas higher-quality diets are associated with greater affluence, energy-dense diets that are nutrient-poor are preferentially consumed by persons of lower socioeconomic status (SES) and of more limited economic means."

Because ethnocultural vegetables are cultural extensions of multicultural Canadians from a variety of ethnicities, the issues of acculturation and ethnicity are importantly related to culturally appropriate food. A potential negative consequence of acculturation is the loss of interest in consuming the healthy vegetables of one's homeland even as other, often better-off Canadians—for whom ethnocultural vegetables were once unfamiliar—begin to consume them (see Chapter 5).

Multiculturalism versus Transnationalism

The term *transnationalism* has received more attention within sociological and geographical circles recently. Vertovec (1999, 447) has described transnationalism as the "multiple ties and interactions linking people or institutions across the borders of nation-states." Similarly, Crang et al. (2003, 439) say that transnationalism means that cultural decoupling and recoupling are occurring such that "cultural identities are no longer clearly wedded to particular nation states, and places are rethought not as intrinsically bounded entities but as constellations of connections within those wider cultural circuits." They believe that transnational geographies can provide a corrective to the cultural globalization that accompanies displacement.

Our understanding of Canadian ethnocultural relations, multiculturalism, and acculturation is very similar to Kymlicka's (1995) and Canada's 1971 policy of multiculturalism within a bilingual framework. While we accept that new immigrants must learn either English or French in order to integrate, we believe, as does Kymlicka (1995), that they should be allowed to keep their cultural practices, including their culturally preferred food. We agree with Jedwab (2016, A13), who points out that "Multiculturalism has led to higher rates of naturalization than ever before. With no pressure to assimilate and give up their culture, immigrants freely choose their new citizenship because they want to be Canadians."

This perspective contrasts with that of Bissoondath (1994), who considers multiculturalism to be a divisive, even potentially disintegrating concept. By contrast, we accept Kymlicka's concept of multinational federalism and the cross-cultural psychological view that acculturation arises via transcultural and cross-cultural interactions, which include the maintenance of people's ethnic traditions and access to their ethnocultural food.

A recent study by Damböck (2009) about Canadian immigrant writers, especially from India, discusses literature dealing with migration, transcultural, and transnational themes. Damböck points to what she considers are differences between multiculturalism and interculturalism, arguing that the social realities accompanying mass migrations increasingly have generated identities that are both transcultural and transnational. She approvingly cites Bissoondath (1994), who believes multiculturalism erroneously regards particular ethnic groups as distinct cultures, which are therefore separable. This, he argues, ignores their evolving transcultural identities (Damböck 2009). But the concept of multiculturalism eschews the American notion of a mosaic or a melting pot, nor does it advocate a static kaleidoscope of cultures. Instead, multiculturalism supports a concept of multi-cultures that deserve respect even though multiculturalism acknowledges that cultural or ethnic groups inevitably undergo change as they evolve in contact with other cultures. Damböck (2009), on the other hand, contends that the term *transculturalism* better describes cultural interaction than does the term multiculturalism because the former acknowledges that the contact from cultural interaction alters both cultures: "Furthermore, due to changing patterns of migration, globalized communities, especially in today's urban centres of the West can no longer be described as multicultural, but are transnational in their composition" (Damböck 2009, 173–74).

There are two problems with her formulation. The concept of multiculturalism does actually imply the acceptance of cultural differences, including people's linguistic rights within their communities, and it acknowledges dialogical cultural interaction while accepting cultural transformations within interacting ethnic and national groups. Multiculturalism does not require assimilation or homogenization. In this way it does not differ substantively from interculturalism. Within Canada, most recent immigrants are therefore not genuinely *transnational* groups. Thus Damböck misses the distinction between ethnic and national groups. In this sense the term *ethnocultural vegetables* is a more appropriate and more general term than *national cultural vegetables* for it refers to the culturally preferred vegetables of all ethnic groups as distinct from the vegetables formerly preferred by Indigenous, French, and English Canadians.

Political Economy of Food and Ethnocultural Vegetable Value Chains

Our political economic analysis of the prospects for ethnocultural vegetables in Ontario accepts the dialectical principle that there is a law of social transformation of quantity into quality as incremental quantitative changes eventually lead to social changes and on rare occasions to revolutionary transformation. Dialectical systems thinking involves identifying those things, classes, or individuals who are contradictory even in their unity (Cornforth 1961). However, social contradictions are often displaced via complementary elements between and among systems for periods of time, but those contradictions inevitably lead to social and natural changes because being or existence is always constituted by becoming (Harvey 1996). While it is true that the presence or absence of ethnocultural vegetables may very well not embody the central contradictions of Canadian capitalist society, the future of Canada is even more multi-ethnic and multi-racial, so multi-ethnic sovereignty over these vegetables will lead to their proliferation whether imported or increasingly homegrown.

Our political economy of ethnocultural vegetables utilizes an historical class analytic approach, which identifies combinations of evolving modes of production. Friedmann and McMichael's (1989) food regime theory further helps us to situate international food relations prior to the present conjuncture.

In 2005, Friedmann summarized the work that she and McMichael had done identifying food regimes since the first settler-colonial food regime from 1870 to 1914 accompanying the rise of industrial capitalism at the behest of the British Empire.

> The Settler-Colonial food regime thus unfolded via three mutually reinforcing effects of government policy: emigration from Europe, settlement of lands converted from indigenous use to commodity production of European staple food, and long-distance shipment of low priced wheat and meat. (2005b, 127)

This was followed, after a long crisis encompassing the Depression and Second World War, by the Mercantile-Industrial Food Regime, which was led by the United States. They subsidized wheat and livestock and increasingly shifted to an export-led economy based on corn and soybeans, which in turn were fed to animals and processed into the kind of pseudo-food that make up what Winson (2013) calls the "industrial diet."

As the latest international corporate food regime became increasingly global, standardized, homogenized, and fossil-fuel dependent, counter-consumer trends emerged that are demanding what the North American food system has largely failed to provide: high-quality food that is traceable, safe to consume, locally sourced, organic, authentic, fresh, sustainably produced, and culturally appropriate (Donald 2009). Largely to counter this corporate food regime's perceived industrial diet and terrible environmental record, alternative, food sovereignty and local food movements have arisen (McMichael 2005).

While questioning whether a third food regime has yet developed, Bernstein (2016) doubts whether agro-ecology, small farmers, and peasants can really produce amounts of food equivalent to the present agro-industrial system. Those favouring the view "that food insecurity is primarily the result of a lack of supply have also opted for variations of this 'just produce and/or import more food from somewhere' strategy" (Wittman et al. 2010, 3).

Consistent with McMichael's perspective, Wittman et al. (2010, 2) define food sovereignty "as the right of nations and peoples to control their own food systems, including their own markets, production modes, food cultures and environments." They believe that the commonly accepted definition of food security has focused too much on increasing total food production "without particular attention to how, where and by whom food is produced" (2010, 3). Thus the food security focus has been uncritical of the corporate food regime's methods of production, distribution, and consumption.

While Bernstein (2014, 1034) has acknowledged that a lack of food security is generally not due to a shortfall in production but is instead due to the "relations of distribution (who gets what)," he notes that "The difference between *buying* food and *producing* it for self-consumption is often elided, however (with a strong preference of Food Sovereignty for the latter)."

Food sovereignty is also closely connected with people's local cultures. It is associated closely with "local production for local consumption" (Wittman et al. 2010, 8). This places it squarely within the local food movement. On the other hand, the notion of food sovereignty is opposed to the commodification of food, such as it happens with the production of agro-fuels, which in turn poses a threat to food security (Wittman et al. 2010).

McMichael (2016, 650) counters Bernstein with the argument that food regime analysis is vital to the task of examining "the political and economic (and now ecological) attending the production and circulation of food on a world scale," though he acknowledges that other analytical issues such as gender, labour, race, ethnicity, and diet must be combined with food regime analysis.

Our industrial diet has been and continues to be nutritionally degraded due to the ever-growing abundance of "pseudo-food," an even broader category than the more widely used term "junk food" (Winson 2013). In 2011, *The Global City: Newcomer Health in Toronto 2011* reported that, while immigrants usually arrive in a healthy state, they may eventually find that their health declines, especially if they are poorer, female, and/or of atypical sexual orientations. Recent immigrants' decline in health status over time is probably partly due to their diminishing access to the kinds of traditional food they were used to consuming before they came to Canada. This outcome occurs because of the kind of acculturation they often undergo, transitioning away from consuming ethnocultural vegetables and other culturally preferred food to eating more of the "industrial diet."

Consolidation within the fruits and vegetables industry has increased the number of "logistical activities performed directly by the producers, such as packaging, distribution and final delivery of the products to the customers" (Ahumada and Villalobos 2009, 3), and it has also been a factor inducing more vertical integration from production to consumption.

On the production side, as part of the process of the penetration of *capital*, family farming operations and small, independent firms are frequently being taken over by larger firms with a more tightly aligned production and marketing chain. Farmers are told to "get big or get out," so there has been a collapse of mid-level farming throughout much of Ontario (Filson and Adekunle 2011) and this is also true within horticulture. There has also been an increasing tendency for retailers to bypass wholesalers and purchase fruits and vegetables directly from the farmers.

Some supermarkets are undergoing certification as a regulatory way to differentiate themselves from others, implement a set of standards, and reduce their transaction costs (Hatanaka, Bain, and Busch 2005). Certification is being done by farmers of sufficient size to be seen as either sustainable or local. In Ontario, for instance, many farmers have also been certified as producing local food with the help of non-profits like Local Food Plus (Bell 2013) and The Stop. The Stop is a west Toronto community organization that began as one of Canada's early food banks and now operates as a community hub providing healthy food and supportive community connections (Van Halem 2012)

Supply chain governance occurs along the entire value chain as people communicate with each other to make decisions about inputs, activities, and outputs. Producers and exporters must communicate with intermediaries, who are often independent importers for supermarkets. Sometimes when the producers are large and technologically advanced, they are in a position to control the relationships and these are producer-driven chains.

More often, the chains are buyer-driven by the supermarkets and super-centres themselves who possess more capital and technology than the producers (Dolan and Humphrey 2000). Gereffi et al. (1994) claim that almost all agricultural commodity chains are now buyer-driven "with both horizontal concentration in trading, and a shift in control downstream from wholesaling to retail" (Raikes and Gibbon 2000, 55).

The Canadian Federation of Independent Grocers is concerned that Loblaw, Sobey's (which is owned by Empire Col. Ltd.), and Metro Inc. make up more than 70 percent of all food retail sales in Canada (Strauss 2014). In March 2014, the Federation asked that the federal government create a code of conduct, as it has for wireless companies, to be overseen by the Competition Bureau. Meanwhile, the Retail Council of Canada, which represents the big retailers, doesn't believe that such a code is necessary. Thus a contradiction also exists within the retail sector between the big "Bay Street" players and the small retailers on "Main Street."

By contrast, there has been a faster concentration of buyers, and the extent of their control over fruit and vegetable retailing is very worrisome. The growth of more individualistic tastes in fruits and vegetables and high-technology monitoring of point of sales has given the more concentrated retailers, compared to farmers and wholesalers, better control and understanding of shifting consumer behaviour. Nevertheless, there is also considerable concentration of commercial fruit and vegetable production on the producer side as well. The larger Ontario producers usually employ seasonal guest workers from the Caribbean and Latin America, and some of the smaller operations do as well. The commercial growers produce most of the fruits and vegetables consumed in the province except during the winter, when the produce available is largely of European/American origin. The much smaller, recent immigrant farmers producing most locally grown ethnocultural vegetables mainly work at non-farm jobs during the winter. The smallholder producers are disadvantaged by their relative lack of access to expensive land, inability to guarantee a sufficient volume of product, as well as the unreliability of their products' quality. Small-operation farmers do not barcode their vegetables, nor can they guarantee high-quality trimming and packaging like the commercial producers (Raikes and Gibbon 2000).

Each type of food has its system of provision, including various connections between production and consumption (Fine 1994), and as with ethnocultural vegetables in Canada, certain vegetables like okra have several systems of provision depending on whether they are imported, produced by domestic commercial farmers, or grown on small farms or in a community or backyard garden. Different types of consumers from individuals to public institutions, including universities, senior citizens' homes, school nutrition programs, and other government institutions, depend on different systems of provision, so that, for example, these varying systems of provision may be most suitable for particular individuals or collective groups. These political economic perspectives on food provision enable discrete class analysis and the determination of the complexity and organization of dominant and subordinate value chains (Raikes, Jensen, and Ponte 2000).

As supermarkets metamorphose from retailers into providers of capital (Burch and Lawrence 2009), the long-distance agro-food value chains are being restructured as supermarkets source fruits and vegetables internationally. There is now almost oligopolistic control of the corporate food regime by large food empires (van der Ploeg 2010). Purely retail supermarkets have become increasingly financialized, strengthening their control of ethnocultural vegetable producers in less-developed countries while controlling most sales in Canada. This financialization involves supermarket chains' control of both land and other assets in semi-tropical and tropical countries that produce ethnocultural vegetables. There has also been a swift expansion of ethnic stores, which usually compete more with the no-frills than the high-end supermarkets.

In order to understand the contradictions that generate changes within the social and technical relations of ethnocultural vegetable production and distribution, it is necessary to clarify some discrete social class characteristics of people within these vegetable value chains. Instead of using the relatively useless gradational class categories of upper, middle, and lower, neo-Marxist categories are used here based more on Wright (1978, 1985) than on Poulantzas (1975) or Carchedi (1977) (see Filson 1983).

Class categories follow from the dominant mode of production within any particular social structure such as Canada's. The point of departure for defining discrete social classes within this political economic framework is one's relationship to the means of production within the particular sector of the ethnocultural vegetable value chain. Owners of those means of production (*capital*) are commercial farmers (*capitalist* horticultural producers) or, for example, owners of supermarkets (e.g., Galen Weston is a major capitalist owner of Loblaw), and they also control investment within their segment of the ethnocultural vegetable value chain. *Workers* neither control investment nor how they must work; instead, they sell their labour for a wage. Between owners and direct producers there may be *manager/technocrats* as well as *supervisors* overseeing production (Filson 2012).

The interaction among these discrete social classes helps us understand the ethical dilemma of Canada's market-oriented local production of food, which involves the use of a particular class segment of the proletariat, seasonal agricultural *workers* who toil at or even below the minimum wage. Also, because they usually have substandard housing, they are not free to move to work for other employers and do not have access to the same health care provisions as do other workers; they are not fairly treated in

Canada, even though they regularly earn more than they would in their home countries.

Balkissoon (2016, A13) points out that "Farm labourers in Ontario, including SAWP migrants, are exempt from labour laws that govern minimum wage, overtime and rest periods," and they are not permitted to have their families join them in Canada. Not only are they not allowed to become citizens, even after working here for as many as thirty years, they have "no pension and no guarantee of health care here after an injury."

"In 2001, the Supreme Court of Canada ruled that the failure to protect agricultural workers' rights violated the Charter of Rights" (Grisdale 2016, A11). Even though the Ontario government passed the Agricultural Employee Protection Act, "no union has been able to strike a collective bargaining agreement" using this legislation (Grisdale 2016, A11). And "there's nothing ethical about 'buying local' and supporting small farms when the workers on them are being exploited" (Gray 2016).

Local Food Movement

Instead of sourcing food through the corporate food regime as is usual for supermarkets and super-centres, local food markets usually mobilize junior value/supply chains supplied mainly by small-scale, local family farmers. The alternative supply chains for this counter-consumer trend still remains marginal compared with the long-distance supply chains of the dominant industrial food model. Despite this apparent marginalization, the counter-consumer trend and community of food practice has momentum, especially in Europe and North America, whether described as the local food movement (Hinrichs 2003; Local Food 2013), the food sovereignty movement, or the community food security movement (Baker 2004).

Donald (2009) references a *Specialty Food Report*, which claims that Ontario's creative food sector (ethnic, local, organic, and specialty food) has grown by between 15 and 25 percent in the last decade. But as fast as new ethnic stores and ethnocultural vegetables appear, supermarkets and super-centres respond with expanded ethnic fruit and vegetable shelf space in their food-scapes. The ethnocultural vegetable market is diverse due to the nature of the demand for these vegetables and the existence of many actors in the value chain. Chain supermarkets and ethnic grocery stores are important in ethnocultural vegetable market-ing. Retailers obtain these fruits and vegetables from different sources—by importing them directly, using intermediaries to import them, obtaining them from the Ontario Food Terminal, or sourcing them locally—but there are only limited volumes of local produce, mainly in summer months.

The large Canadian supermarkets like Loblaw, Metro, and Sobeys have broken the "link between sustainable and local implicit in the original organics social movement" by bringing in organic food, for instance, grown in California (Friedmann 2007, 391). Smith and MacKinnon (2007) defined "local food" as food produced within a 160 kilometre limit, but the Canadian Food Inspection Agency (CFIA), which enforces food labelling in Canada, had a 50 kilometre limit until May 2013, when the CFIA expanded its definition of "local food" to include food pro-duced within each province or territory and also within 50 kilo-metres beyond that. Most provincial producers would probably agree with Local Food Plus's view that local means Ontario grown (Porter 2013, A4).

Some of the local producers are actually growing ethnocultural vegetables in community gardens like Brampton's McVean Farm, which has been providing an opportunity for people of many dif-ferent ethnicities to grow their vegetables. These are rarely prof-itable farms like most urban market gardens, though they have provided the function of being incubator farms because the most enterprising of the McVean farmers have actually transitioned into renting and eventually owning acreages in Vaughn (such as Judy's Tropical Gardens) and other urban areas surrounding the Greater Toronto Area. For the most part, however, the price of land remains beyond the means of recent immigrants so, to the extent that they are involved with commercial production, they are either Canadian or exploited seasonal workers.

The counter-trends to the dominant food regime have been generated by demands for a restructuring of regulations affecting food chains for some time (Buttel 2001). Opting instead for shorter value chains, many consumers have focused their eating on quality local food that is produced in harmony with nature (Higgens, Dibden, and Cocklin 2008). However, there are few opportunities for consumers to obtain sustainably produced food unless they purchase their food in farmers' markets, community shared agriculture, or buy directly from farmers (Friedmann 2007).

Local Food Plus has certified local producers of food since 2005 if they comply with a set of standards regarding the use of labour and local production. Foodland Ontario, in existence for thirty-five years, supports Ontario-grown organic food and is sponsored by Ontario's provincial government. It also uses media outlets, including social media, to promote Ontario-grown fresh food (Foodland Ontario 2013). There are a number of other organizations that encourage more local production in Ontario and that also often promote organic food production (Bell 2013). The *locavore movement*, which advocates production and consumption of local food, started in 2005 (Stroll's 2013), though there have been local food movements in the past.

No doubt some of the imported ethnocultural vegetables may be more affordable despite the extra food miles that are travelled, but is that true for highly perishable food like fruits and vegetables? Can imported vegetables really be that much cheaper than locally produced food, which often has the advantages of being produced with significantly fewer greenhouse gas emissions and higher nutritional value due to greater freshness? Transportation costs are usually not a major part of the cost of growing many vegetables, and part of the competitive advantage of those countries is their relatively low cost of labour. Furthermore, if certain countries have a competitive advantage, they can produce a lot more of a particular produce and benefit from capturing returns to scale. By contrast, Iowa State University's Leopold Centre for Sustainable Agriculture argues that money spent within a community will improve that community's overall income and economic activity, producing new jobs and better profits for farmers, but they caution about the economics of producing most of a community's food locally (DeWeerdt 2009).

Feagan (2007) also believes that "shortening food chains" and encouraging the purchase of local food can improve communities' social and economic viability. Studies of the economic benefits of local food systems are uneven, however, and little research has been done on the net benefits of local food markets (Martinez et al. 2010, 45). Saunders and Hayes (2007, 48) argue that "The shorter distance traveled in local markets was offset by the greater transportation efficiency of the mainstream system." Clearly, as they acknowledge, there's a trade-off between fresher fruits and vegetables transported by air and the extra food miles, CO_2 emissions, and energy use they incur.

Life cycle assessment (LCA) considers the ecological impact of the commodity's life cycle, from its origin to the finished commodity. Cheaper labour costs and better growing conditions explain why this food can be imported. For instance, LCA has shown that it is still cheaper to grow tomatoes in Spain, then transport and sell them in the U.K. (Desrochers and Shimizu 2012). Generally speaking, however, produce imported by air produces relatively more greenhouse gas. Thus for *locavores*, "localization provides the antidote to globalization" (Hinrichs 2003, 34), though she observes that systems analyses of food and agriculture reveal a more nuanced interrelationship between global and local food provision, including cases where good food is unavailable.

Spatial Inequity and Food Deserts versus Pricey Fruits and Vegetables

"In the 1990s the idea of food deserts came into vogue in Britain to describe areas where residents had poor access to affordable and healthy foods" (Black and Macinko 2008, 6). They define food deserts "as places that lack local supermarkets and fresh foods, have unaffordable prices for healthy foods, are socioeconomically deprived and where residents are dependent on corner stores to purchase food" (2008, 6). Since then more North American attention has been paid to food desert locations within rural or urban areas where there is a paucity of supermarkets from which people can access fresh fruits and vegetables and low-fat food items at reasonable prices (Schafft, Jensen, and Hinrichs 2009; Sharkey and Horel 2008; Larsen and Gilliland 2008). Food deserts are usually places of poverty where there is also considerable childhood obesity and often more relatively recent immigrants.

Pearson et al. (2005) found that vegetable consumption tends to go up with age, though men tend to eat fewer fruits than women. These demographic factors are often more significant than consumers' distance from their nearest supermarket, their relative economic deprivation, and even vegetable and fruit prices. However, they thought that in more deprived areas where there was less car ownership, income and prices are much more salient.

While recent Canadian studies have shown that food deserts are not as much of a problem as in many American locations, Larsen and Gilliland (2008, 3) have shown that "studies in London, Ontario and the nearby Waterloo Region have found that residents will have to pay an average of 1.6 times more for identical food items purchased at area convenience stores versus supermarkets." Grocery store suburbanization has resulted in food deserts within the eastern and central part of London. Larsen and Gilliland observe that in larger Canadian cities like Toronto, Montreal, and Edmonton, supermarkets can still be found and are still being constructed due to the larger downtown densities that these cities provide. According to Apparicio, Cloutier, and Shearmur (2007), Montreal really doesn't have food deserts, though middle-income earners have a shorter distance to travel to stores than others do. Beaulac, Kristjansson, and Cummins (2009) observe that there are half as many grocery stores, while convenience stores are three times more common in Hamilton's lower income areas (Latham and Moffat 2007).

Donald (2009, 4) also argues that "the North American processed food diet is the food of North America's poor and working class. Research has found that the poor and lower working class consume the most processed food with the highest proportion of sugar and salt." For this reason poorer people are more susceptible to obesity, and in turn to diabetes and heart disease. Not surprisingly in 2009, Donald noted that 47 percent of Canadians are obese or overweight.

Spatial inequities also exist within foodscapes. Winson (2004) uses the concept of foodscapes to explore what Burch and Lawrence (2009) refer to as the *rentiers* feature of supermarkets—charging for shelf space. Low-profit items, for instance, tend to be placed at the back of the store. Winson (2013) argues that "pseudo-food," which is high in fat and sugar, is highly profitable and therefore is placed in much more prominent foodscape places such as near the cash registers. Consumption of restaurant food, especially fast food, is also closely associated with obesity (Neff et al. 2009).

Tied as well to unhealthy pseudo-food, which is low in nutrition, is a cheap food policy (Sage 2013). Although writing about the American system, Neff et al. (2009, 288) argue that the same has also been true in the United States because their food system policy has tended to be "driven by financial and political power in the food system, including agribusiness and food processing lobbies." Keeping prices low and increasing the supply of a few crops like soybeans and corn guarantee sufficient food supply and provide processors with an incentive to use these basic commodities as ingredients for processed food, which has high profit margins. Muller et al. (2009, 233) argue that "low-priced corn resulted in the rapid growth in high-fructose corn syrup consumption and a significant increase in per capita sweetener consumption."

Furthermore, because highly nutritious fruits and vegetables are more expensive than junk food, they tend to be consumed more by wealthier people or those ethnic groups for whom these ethnocultural vegetables are their chosen ethnic food. Within food deserts there are often many convenience stores that carry processed food but few fresh fruits and vegetables (Neff et al. (2009).

A 1998 study of recent immigrant populations in Toronto looked at the food choices of these immigrant groups. The participants in the study said that finding fresh, culturally appropriate food was important and that while they could find many of their ethnically appropriate food at ethnic stores and sometimes supermarkets, access to them was at times limited and when they could find them, these vegetables were either too expensive, too far from where they lived, or not fresh enough to buy. There was also a shortage of clear information about their preferred food (Koc and Welsh 2002). The tendency, over the past several decades, of increasing globalization along with free trade deals like the North American Free Trade Agreement (NAFTA) has emphasized greater separation between food producers and consumers as more and more food is imported and exported.

Should Food Preferences Be Included in Canadian Human Rights Codes?

As Ramanujam, Caivano, and Abebe (2015, 19) have observed, "While the concept of food sovereignty overlaps with the human right to food, food sovereignty encompasses elements not yet contained in the international community's legal definition of the right to food."

Within the social class and ethnic contradictions accompanying the growing commodification of food, differential rates of access to ethnocultural vegetables are issues of social justice. The consequences are clear of the relative abdication by Canadian governments of their responsibilities under international human rights law, to which Canada is a signatory, to guarantee all Canadians' food security.

Although related phenomena such as ancestry, colour, race, citizenship, ethnic origin, and place of origin are protected human rights grounds under the Ontario Human Rights Commission, ethnocultural food preferences are not (OHRC 2016). In order for that right to be guaranteed, there would have to be a massive increase in public spending to reduce economic inequality, a transfer of wealth that the rich simply would not allow under capitalism. However, much more can be done in the short run to guarantee people's access to their culturally appropriate food. It's time to include food preferences in Canadian human rights codes.

CHAPTER 2

Greater Toronto Area Preferences for Ethnocultural Vegetables

Bamidele Adekunle, Glen C. Filson,
Sridharan Sethuratnam, and Dario Cidro

New Canadians in the Greater Toronto Area (GTA)

Immigrants are the main source of new growth in Canada, a country of more than thirty-six million, and a substantial percentage of them settle in the GTA, Montreal, and Vancouver. The GTA has more than six million inhabitants, making it the fourth largest city in North America, now larger than Chicago (Statistics Canada 2016). Of the three largest ethnic groups, those whose mother tongue is Chinese include those who speak Cantonese, those who speak another Chinese dialect, and those who speak Mandarin. They now comprise more than 500,000 in the GTA.

The largest GTA ethnic groups are South Asian and include those whose mother tongue is Punjabi, Bengali, Hindi, Sinhalese, Urdu, Tamil, and several others numbering over 800,000. For our purposes, South Asia also includes Bangladesh, the Indian Ocean Territories, Bhutan, India, Maldives, Nepal, Pakistan, and Sri Lanka. While all these ethnic groups have some similarities, there are obviously many cultural differences among them as well.[1]

Immigrants from Afro-Caribbean countries are from Jamaica, Guyana, Trinidad and Tobago, Haiti, Ethiopia, Nigeria, South Africa, Egypt, Kenya, Somalia, and elsewhere; speak many different languages; and number well over 400,000 (Statistics Canada 2006b, 2011).

Ethnic Food Consumption and Perception

Ethnic food, or what we're calling ethnocultural food, includes ethnocultural vegetables, plus fruits, meats, fish, and any other food that is commonly eaten by relatively recent immigrants.

While humans are often viewed as rational beings who make informed decisions to optimize their benefits, food decisions are not always rational (Sobal and Bisogni 2009). This is evident as food serves many purposes beyond nutritional value, including construction of personal identity and pleasure (Rozin 1990). There are many factors that influence food choice, including previous exposure, expectations, economic factors, marketing, education and nutritional knowledge, social interactions, morality and religion, culture, lifestyle, age, food trends, media, and attributes of the individual (Asp 1999; Bäckström, Pirttilä-Backman, and Tuorila 2003; McFarlane and Pliner 1997; Nestle et al. 1998; Rozin 1990; Rozin and Tuorila 1993; Shepherd 1999; Sobal and Bisogni 2009).

Biological factors also play a significant role in food choice as they create a general preference for sugary, high-calorie food and a dislike of bitter food, which may be more likely to contain medicinal elements (Rozin 1990; Rozin and Vollmecke 1986). Biology also contributes to neophobia, the fear of new food and toxins, and neophilia, the intrigue with new food and calorie sources, which often occur simultaneously when omnivores are faced with new food (Rozin 1990; Rozin and Vollmecke 1986). Biological responses to the ingestion of new food allow for learning and creation of food preferences, for example, the avoidance of food that causes sickness (Rozin 1990; Rozin and Vollmecke 1986).

Psychology is another major factor in food preference. This includes individual perception of taste, texture, colour, temperature, and presentation (Asp 1999; Nestle et al. 1998; Rozin and Tuorila 1993). The individual's acceptance of these attributes is a determining factor in perceptions about food (Rozin 1990; Shepherd 1999).

Some of the most predominant aspects of food choice and preference are social factors, including present circumstances, past experiences, and culture (Shepherd 1999). Food consumption patterns can be learned in social, cultural, political, and economic circumstances (Sobal and Bisogni 2009). In these contexts, food is consumed to meet nutritional needs, to promote family and friendship ties, to form identity, for holiday meals, to display religious beliefs, to indicate socio-economic status, to reward or punish, as well as to express creativity (Asp 1999). Culture determines how individuals are exposed to food, methods of meal preparation, and serving (Asp 1999; Rozin 1990; Rozin and Vollmecke 1986). Culture also determines what consumption practices are right or appropriate (Sobal and Bisogni 2009).

A more specific example in the literature of how culture shapes food perception is through gender. Gendered ways of eating in Western societies may influence the adoption of ethnic food, especially vegetables. Women in Western cultures tend to be more concerned than men about food, including food safety, health, and weight (Bäckström, Pirttilä-Backman, and Tuorila 2003; O'Doherty and Holm 1999). In these cultures meat is often perceived as more masculine (Wright, Nancarrow, and Kwok 2001), whereas lighter meals containing vegetables and/or poultry are frequently viewed as feminine (O'Doherty and Holm 1999). Apparently, change is the only variable that is constant in life as seen in the recent preferences in the GTA and Canada as a whole.

Willingness to try new food can be understood by learning from studies on children's preferences. There are multiple factors that influence children's food preferences, such as exposure to new food, parental and sibling preference, parental monitoring of food intake, and the mother's consumption of food during pregnancy and nursing (Skinner et al. 2002). Children tend to eat and prefer sweet and fatty food (Skinner et al. 1998, 2002). They tend not to consume as many fruits and vegetables as adults (Skinner et al. 1998; Wardle et al. 2003), and few have ever consumed legumes (Skinner et al. 2002). However, if children observe other people, especially parents and close family members, consuming new food, they are more likely to try the given food (Wardle et al. 2003). Exposure is a more effective means to increase children's willingness to try new food as it can increase their consumption, discouraging neophobia (Wardle et al. 2003). Food preferences can be learned slowly through conditioning with positive sensory cues (Birch 1999; Rozin 1990; Rozin and Vollmecke 1986). It is also more difficult to create a preference than an aversion (Rozin and Vollmecke 1986).

Neophobia increases in early childhood, then decreases into adulthood, and protects humans from dangerous food that may cause illness or death (Birch 1999). As individuals age, they are more likely to try new food, but elderly people may be more neophobic than other groups (Bäckström, Pirttilä-Backman, Tuorila 2003; McFarlane and Pliner 1997; Skinner et al. 2002). However, since the current food environment is less dangerous, neophobia is arguably less useful than it was in the evolutionary past (Birch 1999; Pelchat and Pliner 1995; Pliner, Pelchat, and Grabski 1993).

Yi, Kanetkar, and Brauer (2015) argue that different strategies are required to promote the consumption of a variety of healthy vegetables. They say that "confidence in meal preparation and food skills are key to a substantial increase in the consumption of various vegetables, regardless of income and socio-economic status" (2015, 7).

Apart from exposure, another more cost-effective way to increase willingness to consume new food is through the provision of information (Martins, Pelchat, and Pliner 1997; Pelchat and Pliner 1995). This information is usually effective only if the food is accessible to individuals over the long term because the impact of nutrients is evident after the consumer has consumed the food over a prolonged period (Martins, Pelchat, and Pliner 1997; McFarlane and Pliner 1997). It is important to understand the factors that influence food perception and how it can be altered since there is more new food available in Canada. Even though many immigrants undergo some degree of acculturation, they still want food from their countries of origin (Koc and Welsh 2002). Consequently, the growing number of immigrants contributes to an increasing demand for ethnocultural food (Adekunle, Filson, and Sethuratnam 2012). This demand has led to the availability of cultural food in ethnic food stores and supermarkets (Koc and Welsh 2002). Nevertheless, research on food consumption and perception lacks information on the globalization of food consumption, how markets are influenced by food perception, and the existence and impact of the crossover effect.

Examining People's Ethnocultural Preferences

To understand the main vegetable preferences of the largest numbers of relatively recent immigrants to Toronto, in 2009–10 we surveyed the vegetable preferences of the abovementioned three largest ethnic groups in the Greater Toronto Area (see also Govindasamy, Van Vranken, and Sciarappa 2007). The demographics of the GTA are changing rapidly and now at least half of Torontonians were born outside Canada. As a consequence, the demand for ethnocultural food will continue to grow. Though some of these vegetables can be grown in Ontario, most of them are currently imported. One important purpose of our survey was to provide scientific evidence that will allow some Ontario farmers to benefit from this ethnocultural vegetable market segment by growing the produce locally.

After the successful completion of the pre-testing, in part developed by talking to horticulturalists, the survey was implemented through ethnic stores and associations selected through systematic and purposive sampling. Twelve research assistants, recruited from the University of Guelph, were representative of the GTA's main ethnic groups. Overall, 750 questionnaires were administered (250 per ethnic group). The data were analyzed using descriptive statistics, principal component analysis, and OLS regression. Before summarizing the survey results, we will review some important historical features of these three main ethnic groups and their cuisines.

Chinese Canadians

Chinese Canadians make up the biggest non-European ethnic group in Canada (Statistics Canada 2006b). Eighty-two percent live in Ontario and British Columbia (Statistics Canada 2006d, 2006e).

In Canada, Cantonese immigrants have long adapted their Guangzhou/Hong Kong cuisines from southern China to Canadian tastes and to available ingredients (Civitello 2011; Hui 2016). There is also a fairly strong southern Chinese Sichuan cuisine influence in the Greater Toronto Area (Chan 2011), but other Chinese cuisines have also proliferated in recent years.

Chinese immigrants have been in Canada at least since the Cariboo gold rush of 1858. Hui (2016, L4) argues that early immigrants came seeking *Gum san* or the Gold Mountain, "the nickname the first Chinese workers—the gold–rush–seekers and the railway builders—gave to this place." Seventeen thousand (mainly male) Chinese immigrants came to Canada until the completion of the national railway in 1885, when their immigration was immediately restricted by an exclusionary immigration law in 1882 and the imposition of a $50 head tax, later increased to $500 in 1903. At the time, the weakened Chinese Qing dynasty had a growing number of starving rural people and was near collapse, so many landless peasants attempted to find work abroad. These male Chinese immigrants were not allowed to bring their wives. Though most "sojourners" originally intended to return to China, this became largely impossible (Chan 2012; Ward 1978). Initially they took menial jobs, but soon set up their own restaurants, laundries, and other catering services for the gold miners (Wong 1977). They were not allowed to do most other types of employment. Many former railway workers moved eastward from the primarily hostile British Columbia, where an anti-Chinese riot occurred in Vancouver in 1887 that killed and injured many Chinese (Adachi 1976).

Today there are Chinese restaurants in almost all towns and cities throughout Canada, especially along the railway's route, even on Fogo Island off the coast of Newfoundland (Hui 2016). Toronto's Chinatown began in 1885 (Chan 2011). Despite the head tax, Chinese immigration continued as the Canadian government collected a whopping $23 million until all Chinese immigration was stopped in 1923 as the result of the Chinese Immigration Act (known unofficially as the Chinese Exclusion Act). This Act was finally repealed in 1947 (Hui 2016).

While there were many Chinese restaurants and laundries throughout western Canada, the Chinese encountered serious racism and were not allowed to become Canadian citizens (Roy 1972). Cho (2010, 55) observes that critics acknowledge "the head tax and exclusion laws indicate a legacy of legal racism whereby the Canadian state defined itself over and against Chinese."

As the history of race riots and race-based legislation such as the head tax illustrates, the Chinese Canadian community has been attacked primarily on the basis of its Chinese-ness. Issues of sexuality, class, and gender—especially evident in the promulgation of the perception of a degenerate bachelor society that took jobs away from upstanding and hard-working white men—were crucial in the targeting of Chinese immigrants (Cho 2010, 41).

Chinese Cuisines

In her description of how early Chinese immigrants brought fake sweet-and-sour pork to Canada, Cho observes, after Williams (1977), that the dominant culture incorporates cultural residual elements, but at a distance from itself. What Williams and, in turn, Cho mean by this notion of the cultural incorporation being at a distance from itself (its original Chinese cultural context) is that the dominant culture Williams spoke about (in this case, the dominant Anglo Canadian culture) incorporates residual elements of Chinese immigrant culture, but not in its genuine South Chinese feudal cuisine form. Instead, important skills and practical relationships are ignored or excluded from the original cuisine such that the dominant culture experiences a residual but incomplete version of that cuisine (see Williams 1977). Cho cites the example of fake *goo lo yok*. Meanwhile there is the suspicion that there is a real Chinese menu available to the Chinese, but not to those members of the gullible dominant culture. "The past of sweet and sour pork inexorably seeps into the present of its consumption; the history of coolie labour echoes in the plasticity of its red-orange sauce and colonizes the present of its consumption" (Cho 2010, 37).

She reminds us about the role of colonialism, such as with the First Food Regime, and imperialism regarding mass displacements of people. Cho notes just how precarious the lives of these migrants were and continue to be even today.

Her reading of how Chinese restaurant menus have changed from that of Chew's 1923 New Dayton, Alberta, menu, which contained no Chinese items whatsoever, instead offering tamales and chili con carne. After the repeal of the Chinese Immigration Act, in the 1950s, small-town Chinese restaurants regularly provided both mainstream Canadian and Chinese food. "The Chinese portion of the menu complements the Canadian one by serving up a highly self-conscious stereotypical Chinese-ness that nonetheless produces anxiety through the mechanical reproduction of the menu" (Cho 2010, 65).

The 1951 Diamond Grill's first menu item had evolved to become chicken chop suey and rice, followed by sweet-and-sour pork, egg foo yong, and rice chicken noodle. In so doing, Cho surmises, the restaurants revealed more about whiteness than Chinese-ness. These dishes usually depend heavily on vegetables with chow mein, including peas, bok choy, bamboo shoots, green peppers, and water chestnuts along with fried egg noodles, pork, beef, and or shrimp (Hui 2016). These restaurants provided many opportunities for Chinese Canadian communities, including, most importantly, jobs and they still do.

Chinese immigration resumed in1967 (Hui 2016) with the introduction of the points system (see below), accelerating again in the 1980s with China's impending takeover of Hong Kong (Boyd and Vickers 2000). The variety of Chinese cuisines has in turn mushroomed. Most Chinese Canadians originally came from Cantonese-speaking Guangdong via the ports of Guangzhou (Chan 2011) and Hong Kong, as well as from Taiwan, so early Chinese cooking resembled rural Guangzhou cooking. Recently more mainland Mandarin-speaking Chinese, as well as Chinese speaking other dialects, have immigrated to Canada (Statistics Canada 2007). Roberts (2002) argues that there are eight main cuisines: Shandong, Guangdong, Sichuan, Hunan, Jingsu, Zhejiang, Fujian, and Anhui, though this is contested by others who say there are either more or fewer "high" cuisines (Wu and Cheung 2013). These cuisines have in turn been influenced by Western cuisines since the sixteenth century, when potatoes, corn, and other Western foods were first imported into China. By then, Chinese cooking was already transnational and multicultural because of widespread trade with neighbouring nations dating back to the fourteenth-century Ming dynasty and earlier.

Though there is a close association between religion and food in Western cultures, Mintz (2013, xvi) says, "it is difficult to find a comparable link in Asia." While food is very important in Han culture,

> That dog, cat, snake or monkey, jellyfish or goose foot webs or chicken feet, may turn up in a meal is cause enough for worry among most Westerners. Compared, for example, to the Jewish rules of kashruth, which make some foods prescriptively edible, and others real abominations; or to the rules for Lent in Catholic belief; or to the fasting rules for Ramadan within Islam, Han cuisine seems marked principally by the relative absence of taboo, or of any heavy food-connected emotionalism. (Mintz 2013, xvii)

In the late nineteenth century, Western Chinese food was loosely based on Cantonese peasant cooking (Wu and Cheung 2013).

Increasing demand from newer Chinese immigrants broadened the demand for a greater variety of food as Cantonese overtook Italian to become Toronto's second most spoken language after English (Chan 2011). With the establishment and development of Chinatowns in Canada's major cities, the last few decades have seen a shift to more traditional Chinese cuisines, where there now is an abundance of Cantonese, Sichuan, Mongolian (sometimes Taiwanese), and Taiwanese restaurants, as well as multicultural food such as what is available in Toronto's Kensington Market on Spadina Avenue from College to Dundas.

In the 1970s in British Columbia, Chinese farmers began cultivating *gai lan* (Chinese broccoli), bok choy, and other vegetables (Gibb and Wittman 2013). By that time Chinese farmers also began to be more established north of Toronto in Holland Marsh, where most Chinese greens are produced (interview with Mary Macdonald, University of Guelph 2012). A large selection of other Chinese vegetables is available in large Chinese supermarkets like T&T (now owned by Loblaw) and a plethora of smaller Chinese ethnic stores like the Lucky Moose Food Mart (Asian Grocery Stores in Toronto 2012). Chinese vegetables that cannot be grown in Canada are imported from China and Chinese-owned Mexican farms either directly or indirectly through the Ontario Food Terminal, the distribution centre for 5.1 million pounds of fresh fruits and vegetables a day in Ontario (ECVOntario 2011).

Chinese Canadians, who are found in the largest numbers in Toronto, Vancouver, and Montreal, have well-established networks and offer substantial market potential as consumers. A distinct characteristic of the Chinese Canadians is their tendency to have strong, positive images of Chinese brands, which greatly affects their purchasing decisions (Chia and Costigan 2006). Many supermarkets have created a special Asian food section, and department store owners tend to include Chinese products and services in their promotion brochures (Lee and Tse 1994). Now consider a brief overview of South Asian Canadian history.

South Asian Canadians

There has been a recent trend of South Asian Canadians moving to locations where they can have a better life and enjoy social amenities while maintaining their cultural heritage. South Asian cultural groups, taken as a whole, comprise the largest related ethnic groups in the Greater Toronto Area (Adekunle, Filson, and

Sethuratnam 2010). As with Chinese cuisines, South Asian cuisines have also been well established in Canada for a long time. While many South Asians are vegetarians, other South Asians also eat goat, lamb, chicken, and fish, though beef consumption among this group is rare due to the extent of Hinduism and to some extent Buddhism's beliefs about the sacredness of cattle. For many Muslim Asians, pork is considered taboo as well.

According to Das Gupta (1994), while there were already quite a few male South Asians in Canada by 1904, there were no South Asian women until after 1920. By 1907 there were 2,623 mostly Sikhs from Punjab, who settled in the Vancouver area (Buchignani 1985). There they encountered increasing hostility from local B.C. residents that was fomented in part by the Asiatic Exclusion League. By 1907 South Asians were prevented from voting and could come to Canada only via "continuous journey" (thereby privileging European workers) (Das Gupta 1994). South Asians also had to have $200 in their possession. "The government persuaded the Canadian Pacific Railway (CPR) not to accept South Asians as passengers from Hong Kong and to stop issuing tickets to them (Law Union of Ontario, 1981)" (Das Gupta 1994, 62).

When the $500 head tax was imposed on Chinese immigrants in 1904, that source of labour began to dry up, so private companies began to spread the word in rural Punjab that jobs were available in Canada. Around 1905, East Indian males started migrating to Canada after hearing about the opportunities in the country from British-India troops who had visited the previous year. They were not allowed to vote until 1947, could do only a limited number of jobs, could live only in certain places, and were not allowed to be full citizens. Many Canadians perceived South Asian culture as backward and barbarous (Aujla 2000).

Much of the British Columbian agitation against Asians was perpetrated by the early twentieth-century Asiatic Exclusion

League, which initiated the Vancouver riot of 1907 by holding a parade that degenerated into mob violence as it swept through the Chinese and Japanese parts of the city. At this point, free immigration gave way to controlled immigration when the Japanese also encountered the infamous head tax and worsening disenfranchisement politically and economically.[2]

In the early 1900s, the Punjabis were not allowed to participate in the electoral process, to serve on a jury, or work in professions and civil service. In 1914, when the *Komagatu Maru* Japanese steamer arrived in Vancouver with 376 Indians aboard, having added Indians from Shanghai, Yokohama, and elsewhere to its original 150 passengers from India, it was clear that they had broken the Canadian rule. This rule, designed to keep Asians out of Canada, stipulated that passengers could disembark only if a ship had not docked at any other port since leaving its original port. The *Komagatu Maru* intended to challenge those exclusionary laws, but Canadian authorities refused to allow its passengers to enter Canada or even receive food, water, and supplies from shore. Though some food and water was smuggled on the ship, it was unable to leave as it had no supplies for its return trip to Asia. Only twenty-four who had earlier obtained resident permits were allowed in to Canada, but the ship was forced to leave with its passengers. Later, nineteen passengers died during a clash with the British after the ship arrived in Calcutta, when the British attempted to force them onto a train bound for Punjab, where many did not want to go (Buchignani 1977).

Despite this treatment, in the 1920s the Canadian government finally approved migration of family members of the Sikh men who had arrived earlier, which led to family reunification. The ban against voting and other restrictions were removed in 1947, and immigration of South Asians based on quota was introduced in 1951 (Buchignani 1985). Since then, the number of South Asians in Canada has continued to increase because of education, economic migration based on the points system, family sponsorship, and acceptance of refugees, especially the Tamils from Sri Lanka who were escaping civil war.

The so-called "colour-blind" points system was introduced in 1967 after the White Paper on Immigration. This system gave potential immigrants points based on their education and employability in Canada. This was followed by a Green Paper on Immigration, which added refugee and family reunification classes in 1976. Over the next two decades, a preference for entrepreneurs improved the latter's access to Canada (Naidoo 2003).

Basran quotes Chadeny (1986, 62) as saying that until 1944, 96.4 percent of East Indians were manual labourers, whereas after 1944, 46.3 percent were professionals. Also, until the 1960s, most East Indians were Sikhs (Basran 1993). Basran (1993, 342) observes that "In 1952, a quota of 150 for India, 100 for Pakistan and 50 for Ceylon (Sri Lanka) was imposed to restrict immigrants from these countries." Five years later the number from India was raised to three hundred (Basran 1993), though many Asians faced ongoing racism (Bannerji 1993).

Sri Lanka's conflict between its predominant Sinhalese population and its minority Tamils from 1983 until the government declared victory in 2009 led to an outflow of Tamil refugees and emigrants. Estimates of the number of Tamils in Canada vary from as low as 200,000 to a high of 400,000, though most are Sri Lankan, not Indian, and a sizable percentage live in the GTA (Szczepanski 2015; Hyndman 2003; Sriskandarajah 2002).

Within South Asian Canadian communities, arranged marriages are still widely practised, though there is a significant acculturation to different values in second-generation South Asians.

South Asians tend to integrate secular European cultural elements with their culture; however, family and community structure remain male dominated. The [Talbani-Hasanali] study showed that gender roles were maintained through gender segregation, control over social activities of girls and arranged marriage. Interviewees felt that their parents and communities have more stringent rules for female socialization than any other community in Canada. (Talbani and Hasanali 2000, 615)

Over time, however, these stringent rules have declined as new Canadian-born generations adjust to predominant values.

South Asian Cuisines

Indian restaurants are almost as widespread as Chinese restaurants throughout Canada and their cuisines are as diverse. Nonetheless there is a bit of a struggle between elder Indians' desire for traditional Indian food and their adolescents' desire for at least some Western food. There has been a proliferation of other South Asian restaurants and ethnocultural vegetables in retail ethnic and supermarket stores as well.

Consider how their diets often change over time. Chapman, Ristovski-Slijepcevic, and Beagan (2011, 102) claim that "South Asian Canadians consumed less total fat than Canadians of European, Chinese and Aboriginal origins, but higher amounts of sugar and total carbohydrate." As South Asians acculturate, they often consume a less healthy diet than they did in their home countries, which leads to chronic diet-related diseases like type 2 diabetes. In terms of their culturally appropriate food, South Asians consume a lot of okra, eggplant, bitter melon, cauliflower, and other staples defined by their region of origin. Integration, assimilation, and acculturation have also increased the consumption of this food by people outside these cultural groups.

There is a significant overlap between South Asian as well as Afro-Caribbean ethnocultural vegetables (okra and eggplant, for example) in part because so much of their original habitat was tropical or subtropical. But some ethnocultural vegetables are specifically Asian, such as bottle gourd, or African, such as jute and smooth amaranth.

Afro-Caribbean Canadians

As Weir (2014, 2) observes, "food is a marker of cultural identity. It tells us who we are, how we grew up, about our memories and the history we share. Africa is no different in that respect than other world areas." We are focusing on the ethnocultural vegetables themselves instead of the dishes made from them—such as the pounded-yam *iyan* popular from Ghana to Nigeria, Yoruba *egusi* soup made with *Colocynthis citrullus L.* from the *Cucurbitaceae* family, okra, and other vegetables; the *doro wat* and *injera* in Ethiopia; *suqaar*, *shawarma*, and *anjero* from Somalia; and the maize dishes of south and east Africa. Unlike fruit, vegetables generally must be cooked (see, for example, Weir 2014; Pollan 2013; McCann 2009). We have a blog that discusses many ethnocultural vegetables of Chinese, South Asian, and Afro-Caribbean origin and provides some recipes for cooking these vegetables (http:// evcontario2011.blogspot.ca/).

People of Afro-Caribbean descent in the GTA migrated to Canada through different channels, and there are approximately the same numbers of Canadians of African descent as there are from the Caribbean, though many of the latter also are of African descent. Within Canada, Afro-Caribbean Canadians refers to both African and Caribbean immigrants to Canada.

Ten years ago, there were 317,765 Caribbean and Bermudan immigrants and 374,565 African Canadian immigrants, though many more have come over the past decade. A sizable percentage of them live in the GTA (Statistics Canada 2006a, 2006b).

Canadians of African origin first came to New France and the British North American provinces as slaves in the seventeenth and eighteenth centuries, then arrived as fugitive slaves on the Underground Railroad in the Maritimes, southwestern Ontario, and Victoria in the nineteenth century (Palmer 1975, 7).[3]

The last slave in Canada was a child of a female slave who finally ended his servitude in 1834 at the age of twenty-five. In Lower Canada the courts stopped enforcing slavery after 1797, so 1800 was about the last year for slavery there (Pentland 1981). The contribution of the free blacks and slaves who fought to defend Canada against the Americans in the War of 1812–14 was recognized in 1819, when black veterans were given small land grants at Oro Township in Simcoe County (Hill 1977).

By 1850 Upper Canada had about thirty thousand escaped slaves who came to Canadian towns like Wilberforce, Dawn, and Buxton, mostly via the Underground Railroad. When the Americans passed the Fugitive Slave Act in 1850, the trickle of fugitive slaves soon became a flood because the Act enabled slave-catchers to kidnap blacks even though they were free because they were living in the northern part of the United States (see Moverman 2013). The earlier sympathy for blacks and moral superiority over American whites, which many Canadians felt prior to 1850, began to evaporate as black immigration swelled their numbers to fifty thousand in 1861. Many returned to the U.S. industrial north when it was safe to do so after the American Civil War (Hill 1977).

It is difficult to present a clear route to citizenship for Afro-Caribbeans because they arrived in Canada via different means and for different reasons. Furthermore, our classification has a subset of forty-seven countries from sub-Saharan and a few countries from the Caribbean and West Indies. The challenges of analyzing the immigration path of a heterogeneous group notwithstanding, we will do our best to explain this group's integration in Canada. Immigrants from the Caribbean migrated mainly for access to jobs, education, and family reunion (Henry 1998).

Just as Canada was introducing its non-racist points system, Great Britain began to impose more stringent immigration requirements. As a result of both factors, Canada began receiving larger numbers of immigrants from the Caribbean. Moreover, between 1968 and 1973, those who had been admitted as visitors could apply for landed immigrant status during their stay in Canada. Immigration to Canada from the Caribbean reached its peak during the mid-1970s (Henry 1998). Most of the people from the Caribbean are from the English-speaking former British colonies, with the largest numbers from Jamaica, Guyana, Trinidad and Tobago, Barbados, Grenada, and St. Vincent and Grenadines (Henry 1998). Another characteristic of the Caribbean migration is the "double-lap" migrant whose family first migrates to England and then moves to Canada, as happened with Toronto's chief of police, Mark Saunders, whose parents first immigrated to England and then later came to Toronto (Command Officer's Biographies 2015).

Immigrants from sub-Saharan Africa migrated to Canada as students, economic immigrants through the points system, or as refugees. The first set of refugees from Africa migrated to Canada between the 1950s and the 1970s, when there was a significant influx of Middle Eastern and North African Jews. This was encouraged by the 1951 United Nations Convention Relating to the Status of Refugees. Canada became more involved in refugee protection after 1969, when it signed this convention and its protocol, agreeing not to return people to their countries of origin if they had grounds to fear persecution. In 1972–73, due to Idi Amin's expulsion of Ugandan Asians, seven thousand Ismaili Muslims fled to Canada. There was a surge in the number of African immigrants in the 1990s. At this time asylum seekers came to Canada from all over the world, particularly Latin America, eastern Europe, and Africa. A significant number of Somalis, Sudanese, and Ethiopians were admitted to Canada. In 2010, refugees from more than 140 countries, including Africans, were either resettled or were granted asylum in Canada. Furthermore, in 2011, Canada expanded its refugee resettlement programs by 20 percent over three years (Challinor 2011).

Afro-Caribbean Cuisines

As with the Chinese and South Asians, Afro-Caribbeans' cuisines vary whether they are from the Caribbean, west, central, east, the Horn of Africa, or southern Africa. Caribbean food retains many of the African, Indian, European, Chinese, and Amerindian cuisines. (See, for example, http://www.recipe.com/recipes/caribbean/ and http://evcontario2011.blogspot.ca/2014/05/jamaican-style -steamed-fish-and-okra.html.) Their most common ingredients are many African and Caribbean local fruits (mango, guava, oranges, plaintain, coconut, and jackfruit), vegetables (okra, garden egg/eggplant, cassava, yam, beans, bell peppers, sweet potato, and yucca), cereal grains (rice, sorghum, millet, and wheat), fish, and meats (chicken, goat, lamb, cattle, pork, and some wild "bush meat" like antelope). Popular dishes in central and west Africa include *fu fu* and *garri-eba* (based on cassava), and *amala* and pounded yam (both based on yam). Ethiopian and Eritrean cuisine use the flatbread *injera*, made from teff grain combined with vegetables and meat such as chicken in dishes like *doro wat*. In southern and eastern Africa, a maize-based staple is very common. This food is referred to as *ugali* in the east and *nsima* in the south.

Characteristics of Respondents

The results of our analysis indicate that there is a high demand for ethnocultural food in the GTA (Adekunle 2011, 2012, 2013). Most respondents feel that the consumption of these vegetables is healthy and are willing to pay more if vegetables are of better quality. Most of the Chinese respondents were female (61.2 percent). The South Asian and Afro-Caribbean groups were comprised of 59 percent and 57 percent male respondents respectively. All the men (Chinese, South Asian, and Afro-Caribbean) have spent more years in formal education than the women. The average respondent's age was 40.7 years, average household size was 3.6, and the average total monthly income of our respondents was $3,338.32 across the three ethnic groups. The Afro-Caribbean descendants spent the least percentage of their total expenditure on vegetables. Most of the respondents were married and have at least high school education.

The most important reasons for choosing where to purchase vegetables were availability, proximity/location, and price. The respondents also believe that consumption of vegetables will reduce constipation and is a source of good fibre, prevents chronic diseases (for example, cancer, heart problems like high blood pressure, diabetes); contains vitamins, minerals, and proteins; will reduce medical expenses; helps the immune system; and is part of a healthy diet. There was no significant difference in the level of acculturation of the three ethnic groups. Highly preferred vegetables for the Chinese were bok choy/pak choy/baby bok choy (Cantonese), *xiao bai cai/bai chai* (Mandarin), Taiwan bok choy, Chinese broccoli or *gai lan/kai lan/gail on*, Chinese kale, and eggplant. Afro-Caribbeans preferred okra (Lady Finger), African eggplant (garden eggs), and smooth amaranth. South Asians preferred okra, eggplant (aubergine, brinjal; various varieties based on colour and shape identified), and bitter melon/bitter gourd. As indicated, okra is the most preferred vegetable among South Asians and Afro-Caribbeans. According to Daynard (2016) of the Vineland Research and Innovation Centre (2016, 4) more than six million kilograms of okra are imported into Canada every year.

Daynard argues that between 14.2 and 24.9 million pounds of okra are sold in season across Canada, representing up to

$49.7 million in sales across Canada. About 46 percent of these sales come from Ontario, and 36 percent of that comes from the GTA (Daynard 2016). Daynard (2016) estimates that to meet the demand for okra in Ontario alone would require between 1,400 and 2,600 acres of production (Daynard 2016). The yard-long bean market in Canada is valued at $29 million with 23.7 million pounds sold in season. Chinese Canadians are the largest consumers of the yard-long bean, and 66 percent is consumed within Ontario (Daynard 2016). About 21.4 million pounds of the Asian long purple eggplant are sold in season in Canada, estimated at $33.4 million in domestic sales. About 44 percent of this is consumed in Ontario. South Asians consume about 7.4 million pounds in season, followed by the Chinese, who consume 5.1 million pounds (Daynard 2016).

The demand for ethnocultural vegetables is most prominent among newcomer ethnic populations, while it is much less among Canadians of European origins (see crossover effects in Chapter 5). For example, of the $49.7 million sales for okra in Canada, only 7 percent of this is purchased by Canadians of Aboriginal or European descent whose ancestors immigrated earlier (Daynard 2016). However, these Canadians are becoming more open to consuming ethnocultural vegetables. Many ethnic cuisines have been in North America for a long time and are now Westernized. These include Italian, Mexican, and Asian cuisines. Various other types of cuisines are becoming increasingly popular (Lee, Niode, Simmone, and Bruhn 2012). The Mintel report indicates the growing popularity of Caribbean, Japanese, and Thai food, which indicates that consumers are becoming more accustomed to exploring more ethnic flavours. Popular media outlets, an increase in diversity of community members, more ethnic food in restaurants, and an increase in global travel all provide opportunities for people to learn more about a wider variety of food, which they can replicate at home.

Despite the large demand, the availability of fresh ethno-cultural vegetables is quite limited in Canada today. In super-markets and smaller ethnic grocery stores, the produce is almost entirely imported, so freshness is compromised. VRIC's study indicates that most ethnocultural vegetable consumers are satis-fied with the quality (Daynard 2016), but this may be because there are no alternative sources. Paliyath (Adekunle, Filson, and Sethuratnam 2012b) says that much of the nutritional value of many vegetables, especially brassicas like bok choy, is lost within a week of being picked. Both ethnic populations and Canadians of European descent (see Chapter 5) are willing to pay extra for freshness and better quality of produce. The limited availability of ethnocultural vegetables and the demand for freshness and better quality have encouraged many ethnic populations to grow their own vegetables, usually in their backyards (Adekunle, Filson, and Sethuratnam 2010).

The demand estimates in the GTA indicate that South Asians spend approximately $33 million per month, followed by the Chinese at $21 million, and the Afro-Caribbean group at $7 mil-lion as of 2010. We can conclude that there is substantial and growing demand for ethnocultural food. This opportunity can be explored because there is potential to grow some of these vege-tables locally during the spring and summer.

CHAPTER 3

Ethnocultural Vegetable Value Chain Analysis

Yasantha Nawaratne, Glen C. Filson,
and Bamidele Adekunle

A value chain is "a network of horizontally and vertically related companies that jointly aim at/work towards providing products or services to a market" (Trienekens 2011, 59). Though supply chains are similar to value chains, "supply chain management focuses on how value is added throughout the chain (value added can be defined in terms of (high) quality, (low) cost, delivery time, etc." (Trienekens 2011, 59–60). There are both global and local value chains for ethnocultural vegetables that involve not just growing the vegetables but also their post-harvest handling, distribution (including packaging and transporting), marketing, and consumption. Global ethnocultural vegetable value chains are controlled primarily by the large corporate buyers who prescribe the certification standards the vegetable crop producers must meet. These value chains are very time sensitive, especially when global sourcing is occurring, so a considerable degree of coordination is needed within these chains.

The Greater Toronto Area (GTA) has become ever more dependent on food imports in recent years. In 2007 Lister argued that about 60 percent of the GTA's fresh produce was imported from the United States. These imports have increased greenhouse gas emissions as a result of the distance the produce must travel prior to its arrival (Xureb 2005, cited in Puduri and Govindasami 2011).

Growing consolidation within the agro-food sector has also affected the distribution of power in the food industry's value chains (Reardon et al. 2003). Sobeys's 2013 takeover of Safeway in western Canada and Loblaw's purchase of the Chinese T&T stores and Shoppers Drug Mart are typical of this growing concentration within the retail food trade. The rapid emergence of the large supermarket chains in the past decades has wrested control of most food distribution from smaller groceries. The latest food regime employs mass production of relatively cheap food, the largely unseen costs of which impact our health care system (Seccombe 2007; Lister 2007). These often unseen costs include rising obesity, diabetes, and increasing health care costs (Lister 2007; Winson 2013).

Agro-food systems have been shaped by governments, enterprises, markets, migrating populations, buying, selling, cultural aspects of farming, and consumption. These complex relationships occur during relatively stable specific time periods followed by new possibilities and challenges that in turn create food regimes (Nawaratne 2011). Mass merchandisers have introduced new procurement models to the latest food regime to streamline food distribution (Cook and Davis 2000), which has created competitive pressures for conventional food retailers. In response, retail and wholesale consolidation has occurred to achieve efficiencies through mergers and acquisitions. Consolidation has also occurred for shippers and ownership of strategic alliances through vertical integration.

Researching Ethnocultural Vegetable Value Chains

A number of interviewees, representing every function performed, were selected from different stages of the ethnocultural vegetable value chain. Yasantha Nawaratne (2011) conducted interviews on farms, at research stations, in stores, and at the Ontario Food Terminal (OFT) with importer/exporters, wholesalers, and senior managers, while a focus group comprised of ethnocultural vegetable producers was conducted at McVean Farm. Four researchers from the University of Guelph, Vineland Research and Innovation Centre and Simcoe Research Station, and the University's Muck Research Station at Holland Marsh were interviewed. Joel Aitken (2014) conducted further ECVOntario research into both global and local value chains for okra and Chinese or Indian eggplant, focusing in particular on wholesale market availability, especially at the OFT and its four-hundred-stall farmers' market, where he acquired extensive price data and ethnocultural vegetable observations. Aitken interviewed farmers and wholesalers as well as retail produce buyers, and conducted participant observation at the Ontario Food Terminal.

Through this research we learned that there are primarily two types of ethnocultural vegetable value chains that bring the food to market: both local (shorter) and international (longer) chains, although variations of either exist, such as when local ethnic stores import their vegetables from family and friends in their home countries (intermediate value chains). These value chains include producers, wholesalers, some processors, and retailers. Long global ethnocultural vegetable value chains source produce from various countries and distribute it within the Canadian market. These chains are part of the existing dominant fruit and vegetable supply chains, which are controlled by transnational supermarket chains (Burch and Lawrence 2009).

The shorter chains originate from local producers in Ontario. Small-scale producers tend to be new, less experienced than large-scale producers, and produce their vegetables organically during summer months. They market their goods mainly through farmers' markets, community shared agriculture, and direct farm sales responding to consumers' demand for local, high-quality, fresh, organic ethnic food.

Larger, more commercial-scale ethnocultural vegetable farmers have more organized marketing arrangements to supply their vegetables directly to large ethnic supermarkets and/or to the Ontario Food Terminal and its farmers' market. In addition to summer cultivation, some large farm operators have overseas production arrangements in countries such as Mexico that allow them to maintain a year-round supply of ethnocultural vegetables.

An important supply channel is the large number of independent importers and distributors who provide the produce to the GTA, reflecting the diversity and variation of the supply and marketing of these vegetables. Large supermarket chains purchase many products in bulk and have more bargaining power than other actors. However, producers with strong brands, small firms producing high-valued unique products, or firms with a unique geographical advantage also have a relatively strong position.

These independent importers distribute their ethnocultural vegetables to ethnic stores where they have good business, social, and cultural relationships. Power relationships between retailer and supplier are characterized by their mutual dependency on each other, while power relations between retailer and manufacturer vary with particular social contexts. The variability of the vegetable produce and the actors involved in the value chain enables this to occur.

Product movement from producer to consumer is maintained by a range of supporting and logistics activities, regardless of the origin of the product. These services broadly include input supply, research and extension, transportation, storage, process-ing, and marketing. Supporting activities of the value chain can vary, depending on the industry. According to Porter (2002), the activities in the value chain can be divided into primary and sup-port activities. The primary activities of the ethnocultural vege-table value chains are production; various activities done by the suppliers, including production arrangements; logistics involved in shipping; and storage and marketing of the produce. Support activities incorporate procurement, technology development (including research and development), human resource manage-ment, and infrastructure. The role of the support activities is to support the primary activities to achieve desired productivity.

The Ontario Food Terminal importers have a strong position in the fresh vegetable sector because they are well integrated into the system. Most of their businesses have long-term relationships and partnerships with their suppliers and intermediaries, and they remain current with the latest technological and logistical food chain developments. According to Reardon et al. (2003), the involvement of third parties between retailers and suppliers has increased. Some of the intermediaries in the Ontario Food Terminal and large supermarket chains have diversified their businesses and have integrated their operations vertically with farmers locally and abroad.

The freshness and nutritional qualities of many fruits and vegetables are often in doubt because of the length of time between the harvest of the produce and its arrival at stores and markets for purchase by the consumer (DeWeerdt 2009). Aitken (2014) found that it can take up to twelve days for okra from some central American sources to arrive at the Ontario Food Terminal after it has been picked and packaged, whereas locally grown okra can be on store shelves in two days or less, thereby being much fresher. Nonetheless, Aitken found that Nicaraguan okra is pack-aged and sorted better than our local product, but because it is generally flown in, it arrives with the additional cost of much more greenhouse gas production.

Elliott Currie, of the University of Guelph, has worked with horticulturalists at Vineland Research and Innovation Centre, and argues that it is very economical to grow okra in Ontario. With experienced seasonal agricultural workers from Mexico and Jamaica earning an average of $15 an hour, a farmer can produce as much as $90,000 worth of okra with labour, rental, and other input costs of around $30,000 on a 10 acre plot. Also economical is Chinese or Indian long eggplant, which can be grown for roughly half of what it sells for (according to Currie, June 3, 2016). Yard-long beans, bok choy, baby bok choy, Chinese broccoli, and a number of other Chinese greens have been grown successfully in southern Ontario and British Columbia for many years.

Of course, countries like the United States, Mexico, Nicaragua, and Dominican Republic have better agro-climatic conditions for growing such tropical ethnocultural vegetables as bitter melon, okra, eggplant, smooth amaranth, and hot peppers. Additionally, most of these producers use cheaper labour than is available in Canada, so they have lower input costs than Canadian producers. They also have the advantage of being able to use pesticides and chemical fertilizers, which Canadian producers are not yet permitted to use as there are virtually no registered pesticides appropriate for producing these vegetables, so they must be grown organically.

Supermarkets have the majority of the retail market share in Canada (Boothman 2009). Large chains are becoming more and more interested in selling ethnocultural vegetables after they first appeared in ethnic stores. A buyer for Loblaw revealed that the ethnocultural vegetables now available in most chain supermarkets are being given priority shelf space in supermarket foodscapes. This varies based on the location, the number of ethnic customers, and the requirements of the supermarkets. Since these vegetables occupy a market segment that depends in part on continued immigration, this ethnocultural vegetable market has great potential for expansion. This demand has created a competition, as evidenced by Loblaw's acquisition of T&T stores, which are operated similarly to how T&T's former owners managed them—that is, with largely Chinese staff and management. Metro from Montreal also acquired a majority interest in Marché Adonis in October 2011 (Marotte 2011). This retail and wholesale consolidation takes place as competitive efficiency pressures force stores and supermarkets into mergers and acquisitions (Cook and Davis 2000).

Maintaining product standards, quality, and offering the best value for customers is a big part of supermarkets' marketing philosophy. Supermarkets have global purchasing and centralized distribution centres with quality, freshness, and food safety requirements to achieve this (Reardon et al. 2003). Another significant aspect is the development of private standards, certification, and store-brand labels like Loblaw's President's Choice. According to one Loblaw manager, customer expectations have increased in recent years. Customers want fresh, high-quality ethnocultural vegetables with greater choice, better packaging, a better environment for shopping, and more information about the produce. Smaller supermarkets often do not recognize "locally produced" or "organic" as important factors for ethnocultural vegetables. They also don't have a separate organic section for ethnocultural vegetables. However, large-chain supermarkets typically do and they charge a premium for the organic produce. Supermarkets also collect information about the demand patterns for ethnocultural vegetables, especially about what products are demanded by which ethnic groups, and how demand patterns change over time and geographically.

The ethnocultural vegetable value chain in Ontario consists of a number of sub-chains. These chains are made up of short and local chains, certain ethnic and relational chains, and longer, yet dominant, global chains. Most ethnocultural vegetables consumed in Ontario are produced abroad. As an emerging segment of the existing food system, ethnocultural vegetable value chains are connected to social, cultural, economic, health, and environmental aspects of the food system. Most of the farmers who grow ethnocultural vegetables in their backyards, in community gardens, or at FarmStart's McVean Farm are relatively recent immigrants. Within these shorter value chains, these farmers and their networks play an important role in securing information about growing and marketing ethnocultural vegetables. Farmers cultivate these products not only to earn income but to provide healthy food for their families and friends.

The retail chain managers interviewed said that the production and marketing of local ethnocultural vegetables are generally not well organized. There are many reasons for this, including the complexity of understanding the demand and the difficulty in obtaining fresh ethnocultural produce. Most retail chain managers are from traditional national or ethnic groups and are relatively unfamiliar with ethnocultural vegetables. They tend to depend on traditional supply chains from the United States and the Caribbean. There are also not many local producers of these vegetables and, where they exist, they tend not to be able to guarantee a sufficient supply for the larger chains that can meet those supermarkets' strict production requirements. Ethnic stores depend more on their relations with family or friends from their countries of origin to help them procure these vegetables. Their networks include some local producers from their ethnic groups, but too often local vegetables producers, especially if they are not from the same ethnic groups as those stores, are unfamiliar with how to produce those ethnocultural vegetables under Canadian agro-economic conditions. However, increasingly retail chains are employing people from recent immigrant groups who are more familiar with demand for ethnocultural vegetables. That partly explains the existence of many short and long ethnocultural vegetable chains in the market. The services local ethnocultural vegetable farmers provide and the knowledge they possess have helped them to gain their market share despite competing large importers and supermarket chains.

Figure 3.1 outlines the most salient features of the ethnocultural vegetable value chains. As can be seen in this figure, the existing ethnocultural vegetable value chains fall within short (local) and long (international) value chains.

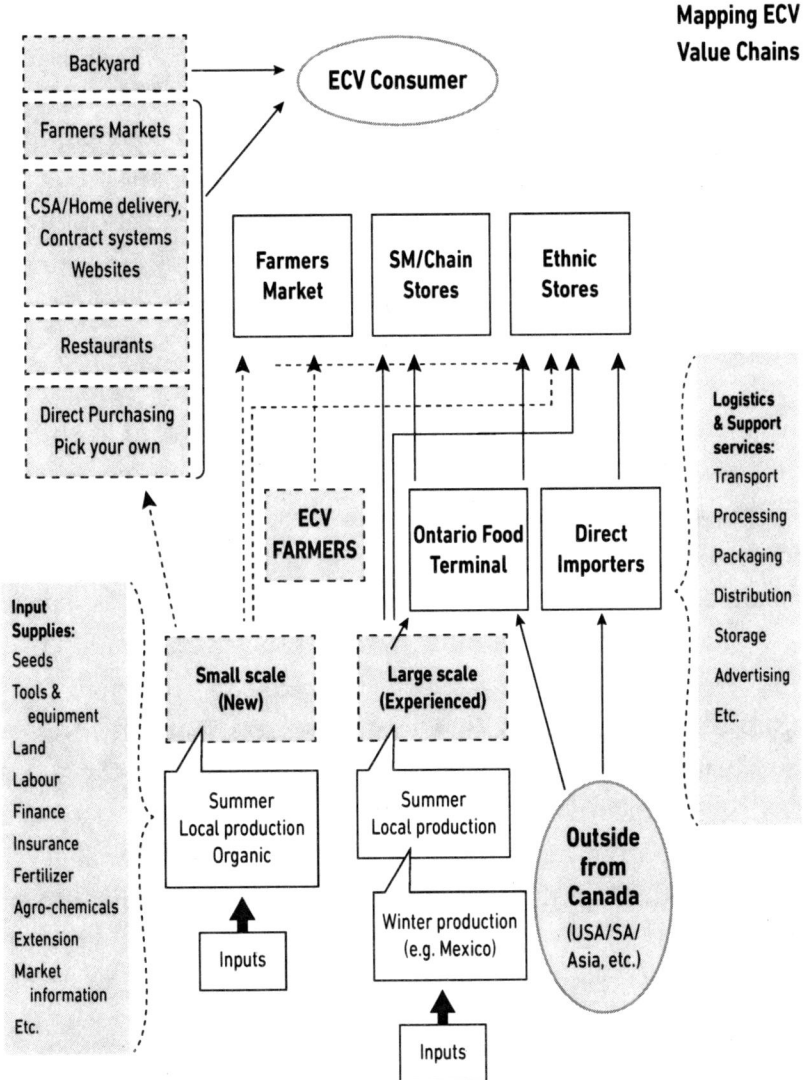

FIGURE 3.1 • Ethnocultural Vegetable Value Chains

Food regime and value chain theories help us understand the changes and various power dynamics in the latest food regime from long- to short-distance value chains. The latest food regime is characterized by the dominant transnational corporations with long-distance food chains that have centralized procurement and product standardization. Supermarkets have secured a strong position in the system, thus enabling them to control the producers, those intermediate elements within the value chains, including procurement, assembly, wholesaling and retailing, and consumers' choices. Horticultural research on ethnocultural vegetables suitable for Ontario soils and conditions is being conducted at Simcoe Research Station (SRS) and Vineland Research and Innovation Centre (VRIC). Understanding the ethnocultural vegetable value chains is critical to the development of appropriate policy interventions required for promoting the cultivation of these vegetables in Ontario.

Shorter food chains create lower greenhouse gas emissions by reducing food miles, though life cycle assessment has shown that this is not always the case (Coley, Howard, and Winter 2009). McMichael (2009) argues that high fossil fuel and energy consumption levels accompanying the long chains will not be sustainable in the future. The negative greenhouse gas consequences of the long, energy-consuming food chains in Ontario's food system are alarming. The emerging public movement against "long food miles" is raising awareness about this lack of sustainability (Friedmann and McNair 2008). Besides the social and environmental welfare gains, Ontario consumers will have the opportunity to taste fresh and healthy ethnocultural vegetables produced locally. From the farmers' viewpoint, apart from the opportunity to cater to the huge demand in the GTA, they can use these vegetables as alternative crops to replace less profitable endeavours.

Ethnocultural Vegetable Imports and Distribution

The Ontario Food Terminal (OFT), established in 1954, is a vegetable and fruit market supplying both ethnic stores and, to some
degree, supermarkets, usually from wholesalers like the OFT's
Ippolitos, who have local suppliers as well as ethnocultural vegetable suppliers in Mexico, the Dominican Republic, the southern
United States, and other parts of the world. The OFT's manager,
Bruce Nicholas, argues that one major advantage of producing
ethnocultural vegetables in Ontario is that it reduces the production cost by as much as 75 percent during the summer, depending
on the vegetables (Nicholas 2012). The OFT is the largest wholesale market in Canada. A very large amount of ethnocultural
vegetables is moved through the OFT to retailers throughout the
year. The OFT provides all logistics and other facilities for smooth
vegetable distribution on a large scale.

The primary intermediaries here are the OFT and the
independent importers. The OFT consists of large-scale importers and wholesalers who import vegetables from many different
countries to reduce the risk of not having vegetables of adequate
quality for sale and maintain consistent supply in response
to growing demand. Many of these giants have their own production arrangements in the United States, Latin America, and
the Caribbean.

The OFT has twenty-one large-scale warehouse tenants
and fifty office tenants who import fresh vegetables and fruits.
The average daily volume of fresh produce traded in this facility
is about 5.1 million pounds per day. The imported produce in
the OFT varies from an average of 65 to 75 percent in which
25–35 percent is locally produced. At the OFT buyers have the
opportunity to see and compare both local and imported vegetables and fruits and make purchasing decisions. Figure 3.2 presents the basic structure of the OFT.

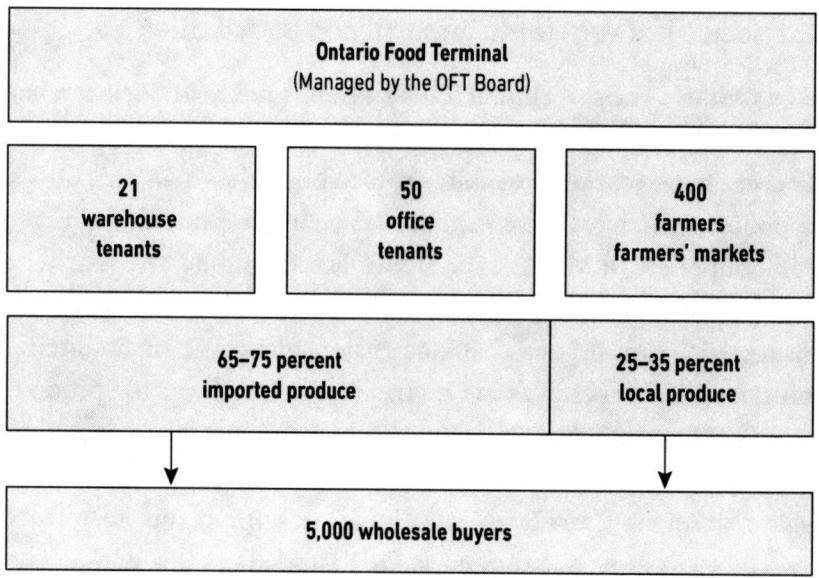

FIGURE 3.2 • Ontario Food Terminal
(Source: General Manager, OFT Board)

There are more than five thousand registered local buyers at the OFT, including supermarket chains and ethnic stores from all over Canada. The OFT is open twenty-four hours a day to registered wholesale buyers, but not to the general public. Most of the ethnocultural vegetables imported to Canada are from the United States, Mexico, Dominican Republic, Nicaragua, Peru, Guatemala, and Colombia. There are also products from Europe and Asia, including China, Korea, and India. Table 3.1 presents some of the common ethnocultural vegetables found at the OFT and their countries of origin.

The OFT also has a vibrant farmers' market, where about four hundred farmers are registered, strictly to market Canadian produce. The local produce, mainly from Ontario and Quebec, generally arrives in the summer months. Ontario farmers have the opportunity to sell their produce in a facility where there are more than five thousand registered wholesale buyers. The number of buyers shopping at the OFT in a day can vary from one thousand to three thousand, depending on the season and the demand.

The OFT Farmers' Market has more mainstream than ethnocultural vegetables, but there is a special section called "Chinese row," where many Asian farmers from Bradford and the Holland Marsh area have been selling mainly Chinese vegetables for many years. All the farmers have the opportunity to sell and compete with the imported produce in terms of price and quality. Local farmers have to display and package the product to compete with international producers, but Ontario farmers have the advantage of selling fresher produce.

One large-scale importer said that ethnocultural vegetables have become increasingly available in the OFT over the past two decades. Currently Chinese supermarkets in the GTA buy a lot of ethnocultural vegetables on a daily basis. According to a senior

TABLE 3.1 • Some ECV Imported Through OFT and Their Countries of Origin

ECV Variety	Country of Origin
Bok choy, baby bok choy, pak choy	U.S., Canada
Napa cabbage	U.S., Canada
Chinese broccoli	U.S., Canada
Okra	U.S., Mexico, Nicaragua
Eggplant	U.S., India, Mexico
Coriander	U.S.
Onion	U.S., Canada
Spinach	U.S.
Bitter gourd	Dominican Republic
Pumpkin	Jamaica
Plantain	Ecuador
Cassava	Costa Rica
Sweet potato	U.S.

supermarket manager, the demographic composition of the buyers has also changed over the years to more closely reflect the GTA's ethnocultural diversity. The buyers during the early years were mainly Italian, whereas now there are more Chinese, Korean, Iraqi, and South Asian OFT buyers.

Ethnocultural Wholesalers and Distributors

Importers' purchasing of ethnocultural vegetables depends on market demand, availability of produce and quantities, cost, consistency of the supply, and the quality. Most importers have a variety of producers from many countries to maintain a consistent year-round supply. This is a well-organized operation between producers, suppliers, and buyers in which trust and mutual understanding are as important as contracts. Most of these business partners have a long history of partnerships.

Some of the wholesalers have integrated backward and forward in the supply chain and diversified their businesses. Some have their own storage, processing, and packing facilities and transportation logistics. A large-scale wholesaler and importer of vegetables in the OFT sources his company's produce both locally and from overseas. This company has been in the vegetable trading business since the 1940s and started at the OFT facility in the late 1990s. Today the business is managed by the fourth generation of the owners' family. The company's general manager explained his company's operations:

> We produce 90 percent of our produce by ourselves. This includes our own farms and arrangements with farmers. The relationship with farmers is very strong. It runs back to many years where they understand our expectations and requirements like what we need and quality. Most of our growers are in Canada. Once we run out of local stuff we get it from the U.S. We started to grow in U.S. where

we have a 2,000 acre farm in California. 10–15 percent of the produce of this farm comes to Canada. This helps us to maintain a year-round supply under the same label. (February 17, 2012)

Another change that has occurred is the growth of the herbs market where, for example, the demand for cilantro (coriander) has increased five or six times compared to ten years ago, while the demand for parsley has declined. The demand for okra, which is imported from the United States, Nicaragua, or Mexico twice a week, is rising.

In addition to the OFT, there are many other independent importers of ethnocultural vegetables in Ontario. They vary in scale of operation; range of product handling; geographical specialization of produce; services offered; level of integration in the industry; and supply of ethnocultural vegetables to ethnic stores, ethnic supermarkets, and supermarket chains.

Ethnic farmers' knowledge and skill are huge assets in producing ethnocultural vegetables and could be used more productively to grow and market more of this produce locally. When more of their production costs are known, more farmers will grow these vegetables. Whereas the Ontario Fruit and Vegetable Growers' Association (OFVGA) is the voice of Ontario's 7,500 fruit, vegetable, and greenhouse farmers regarding issues affecting the horticulture sector (OFVGA 2011), there is no formal association at present to unite or represent their needs. While the leadership of the OFVGA is keen to expand its members' production of ethnocultural vegetables as a new sector that they can take advantage of, the OFVGA does mainly represent vegetable producers who produce traditional vegetables like tomatoes, cucumbers, green beans, etc., in part because this production still represents most of the vegetables produced locally. What is being requested here is a separate organization either within the OFVGA or outside of it that specifically represents those producing ethnocultural vegetables. The majority of the traditional producers probably do not perceive this need despite the huge and growing demand. Such an association should be formed either within the OFVGA or separately.

CHAPTER 4

Consumption of Culturally Appropriate Food:

The Impact of Globalization, Immigration, and the Retail Market Structure

Christine Kajumba, Glen C. Filson, and Bamidele Adekunle

In 2006, in response to rising demand for locally and organically produced ethnic and fusion food, Donald and Blay-Palmer observed that "The extent to which these contested sites of everyday food practice are being, or will be colonized by an urban agro-industrial or food-retailer elite is yet to be determined" (1917). Since then, the largest retailers have moved aggressively to colonize this space, but not without pushback from smaller, ethnic operations.

Ethnocultural vegetables are an important source of healthy food and contribute to the social, economic, and environmental fabric of society. Like other vegetables, this produce ensures more nutrition than processed or fast food (World Health Report 2003). While most vegetables are healthy to eat, some are known as functional food because they are exceptionally nutritious. Functional food is enhanced with bioactive ingredients that have demonstrated health benefits (Herath, Cranfield, and

Henson 2008). Bitter melon is a good example of a functional ethnocultural vegetable. Islam, Jalaluddin, and Hettriarachchy (2010, 61) found that such ingredients "as protein, amino acids, minerals, and polyphenolics contents were determined using four selected varieties" of bitter melon, and that these ingredients do indeed have strong antioxidant qualities and other nutraceutical benefits. Smooth amaranth is another popular functional ethnocultural vegetable (Awolu, Osemeke, and Temilade Ifesan 2016).

Although Canadian immigrants in the GTA have been able to adapt to Canadian lifestyles to some extent, their search for their home food usually remains strong for some time. Other Canadians are learning to appreciate these vegetables for their taste and nutritional qualities (Donald and Blay-Palmer 2006).

There is an increase in nutrition-related chronic diseases. The current epidemic of obesity, cardiovascular diseases, diabetes, and high blood pressure is significantly affected by poor diet, and too many people are not able to meet their nutritional needs (World Health Report 2003). In Canada the cheapest foods available are high in sugar, salt, and fat, and low in nutritional value. Increasing public awareness of nutrition-related chronic diseases has led to a gradual change in food consumption trends, with many people preferring more plant than animal products in their diet (Wolf, Spittler, and Ahern 2005). A diet rich in vegetables reduces the risk of cancer and heart disease (Ziegler 1989), thus prolonging life expectancy.

Vegetables, whether ethnocultural or mainstream like bitter melon and garlic, are known anti-carcinogens and can also affect other diseases (Reynolds-Zayak 2004). Pumpkin seeds, for instance, which are both mainstream and ethnocultural, are known to reduce symptoms of prostate enlargement (World Health Report 2003). As immigrants grapple with the changes in their diets, they are also concerned about their nutrition and health (Koc and Welsh 2002) and, if possible, will continue to eat the ethnocultural vegetables they are used to eating.

Ethnocultural Vegetables, Culture, Faith, and Socialization

Cultural values are variously expressed, including through food (Royce 1982). Food has been used as a symbol of cultural unity, prepared and served at special functions and celebrations, thereby affecting people's attitudes and food choices. Thus food is an intimate part of society and a key factor in one's sense of identity (Fischler 1988; Winson 1993). Food choices, attitudes, and preferences are developed, influenced, and reinforced early in one's life by one's socio-cultural group. Eating one's culturally prescribed food is an ingrained and resilient habit (Rozin and Schiller 1980). Food preferences and familiarity are among the last cultural traits that one loses (McMichael 1995).

Vegetables consumed vary depending on one's origin, and certain vegetables are associated with specific cultural groups (Abdel-Ghany and Sharpe 1997), yet other vegetables may be similar or closely related across cultural groups. In the process of acculturation, immigrants experience change regarding the kind of food eaten and resistance to some new food because of food preferences and lack of familiarity. This may result in either social distancing or integration (Capella and Arnold 1993) as immigrants consume substitute vegetables that are closely related to the food they were originally socialized to prefer.

Food in general, and ethnocultural vegetables in particular, also play important roles as a part of faith and socialization. For instance, vegetarianism is advocated by some Hindu and Buddhist scholars, and we found the highest relative vegetable demand was from South Asians (Chapter 2). People's faith and their preferred food are often closely associated (Serecon Management Consulting 2005).

Jewish kosher food is lawful or permitted food (Masoudi 1993). For Muslims, halal is permitted food, while haram food (like pork) is not permitted. In Hindu culture, eating beef and sometimes pork is prohibited, but eating vegetables is encouraged (Dindyal and Dindyal 2004). These norms can influence individuals' eating habits appreciably (Blackwell, Miniard, and Engel 2001; Delener 1994).

Food has also been used for socialization when it unites people or as an expression of ethnocentrism when it is used to identify "others" (Avakian 1997; Caplan 1997). Specific vegetables may be served at ceremonies as cultural symbols. Different vegetables may signify different cultural practices and are often a compulsory component of a ceremony. Each ethnic group has its culturally distinct cooking styles (Caplan 1997).

Vegetables have also been used as a differentiating element when members of some cultures feel their food is better than those of other cultures. However, many of these cultures and customs have been influenced and modified by constant contact among numerous cultures, modern cooking practices, and lifestyle. With increased exposure and integration of communities, there is greater experimentation with new food as people socialize by sharing meals at home or in "exotic" restaurants. The cultural diversity, coupled with the curiosity and adventurous nature of many consumers, has led to the popularity of and increasing demand for ethnic vegetables in the Canadian market.

Considerable research on food consumption has been done on the influence of age, education, gender, attitude, perception, lifestyle, and the perceived need for variety (Hamlett et al. 2008; Reynolds-Zayak 2004; Morland, Wing, and Diez 2002). More educated Americans tend to eat more vegetables and less animal protein (Glanz et al. 1995), and this is also probably true of more educated Canadians. Granner et al. (2004) found that American women were more likely than American men to select a vegetable-rich diet, and the women were often more conscientious than men about buying healthy food. Similarly, in Canada elderly women are much more likely to select a vegetable diet than their male counterparts (Riediger and Moghadasian 2008).

Older consumers are often more health conscious (Agriculture, Food and Rural Development 2004) and focus on low-sodium, low-fat, highly nutritious, and easy-to-chew food, making vegetables a popular choice. However, their meals depend on the affordability of the food and how easily it can be eaten. By contrast, younger teens are usually less health conscious and eat more processed or ready-to-go food and fewer vegetables, due to their lifestyle, attitudes, and how they perceive food. As the Canadian population gets older, more educated, more ethnically diverse, and have increasingly busy work schedules, lower vegetable consumption becomes a more serious concern.

The Macro Environment Affecting Ethnocultural Vegetables

Globalization has impacted vegetable availability through market liberalization as new powerful players have entered the system, reducing tariffs and liberalizing trade (FAO 2004). This has expanded the importation of ethnocultural vegetables into Canada's market (Zafiriou 2005), thus increasing Canadian access to the produce, offering a greater selection to satisfy personal needs and preferences, and facilitating healthier diets (Zafiriou 2005; Friedland 1994).

During Canada's production season, vegetables are more available both in quantity and quality and sell at a relatively affordable price, allowing for increased consumption. Increased vegetable importation has maintained the stability of their availability, especially during low-production seasons. However, the GTA still imports fresh produce during our most agriculturally productive seasons even though many ethnocultural vegetables such as yard-long beans, okra, smooth amaranth, and bitter gourd can be grown locally (Food and Hunger Action Committee 2000).

The effects of globalization have influenced ethnocultural vegetable distribution chains as global mass supply/importation of vegetables by multinational corporations has grown with increased international trade (Swinnen and Maertens 2007; Goldman, Krider, and Ramaswami 1999). Liberalization of trade has triggered escalated foreign investment in agribusiness, food industries, and food distribution and retailing (Swinnen 2005). In turn, market strategies have changed as supermarkets increasingly dominate the food retail market (Reardon and Swinnen 2004). Importation of cheap vegetables through the global trade system has reduced the prices of many locally grown vegetables (Xuerebo and Desjardins 2005) and enabled cheap vegetables to penetrate the market, but it has also allowed for increased consumption. Food costs have been lowered as a result of increased competition, greater concentration, and specialization of production and distribution (Raikes and Gibbon 2000; Wormsbecker 2007) as more players have joined the global food market.

To some extent, the quality and safety of food has improved due to increased consumer demand (Maertens and Swinnen 2007) and competition as a result of globalization of supply chains and increased pressures to regulate the standards of their products (Wilson and Abiola 2003).

The Micro Environment Affecting Ethnocultural Vegetables

The micro-environment includes the accessibility of ethnocultural vegetables from the nearest grocery store, what one can afford, and the appropriateness, adequacy, and nutritional level of the vegetables. One's proximity to or distance from a grocery store may determine whether or not one purchases from that store, especially if the person depends on public transport (Bodor et al. 2007).

Access to a supermarket also plays a role in consumption of vegetables (Morland, Wing, and Diez 2002; Rose and Richards 2004; Laraia et al. 2004). Also, vegetable availability improves consumers' access to ethnocultural vegetables. If the store offers most of the vegetables, this influences one's chances of purchasing them from that store (Jago et al. 2007). The appropriateness, quality, and nutritional level of vegetables sold are other important factors (FAO 2004) as consumers are specific in their demands. Stores must stock the appropriate vegetables for the right group, and the vegetables need to be fresh and nutritious if consumers are to buy them and be likely to return to the store.

One's social class (Steele et al. 1991; Shimakawa et al. 1994; Roos et al. 2008) and the factors that lead to unequal wealth distribution influence people's vegetable selection. Relatively wealthier consumers are more able than poorer people to purchase more vegetables, especially if organically grown (Regmi et al. 2001), which disproportionately excludes recent immigrants (Neff et al. 2009).

Figure 4.1 sums up the factors that affect ethnocultural vegetables consumption in the GTA. Three main factors affect vegetable consumption: personal characteristics, the macro-environment, and the micro-environment. Among the personal characteristics are lifestyle, health concerns, demographics, and ethnicity. Macro-environmental factors include globalization, immigration, supply chains, and season of production, while micro-environment includes the distance from the nearest store, and the appropriateness and adequacy of ethnocultural vegetables. These factors eventually affect the availability, affordability, accessibility, and finally the utilization of ethnocultural vegetables.

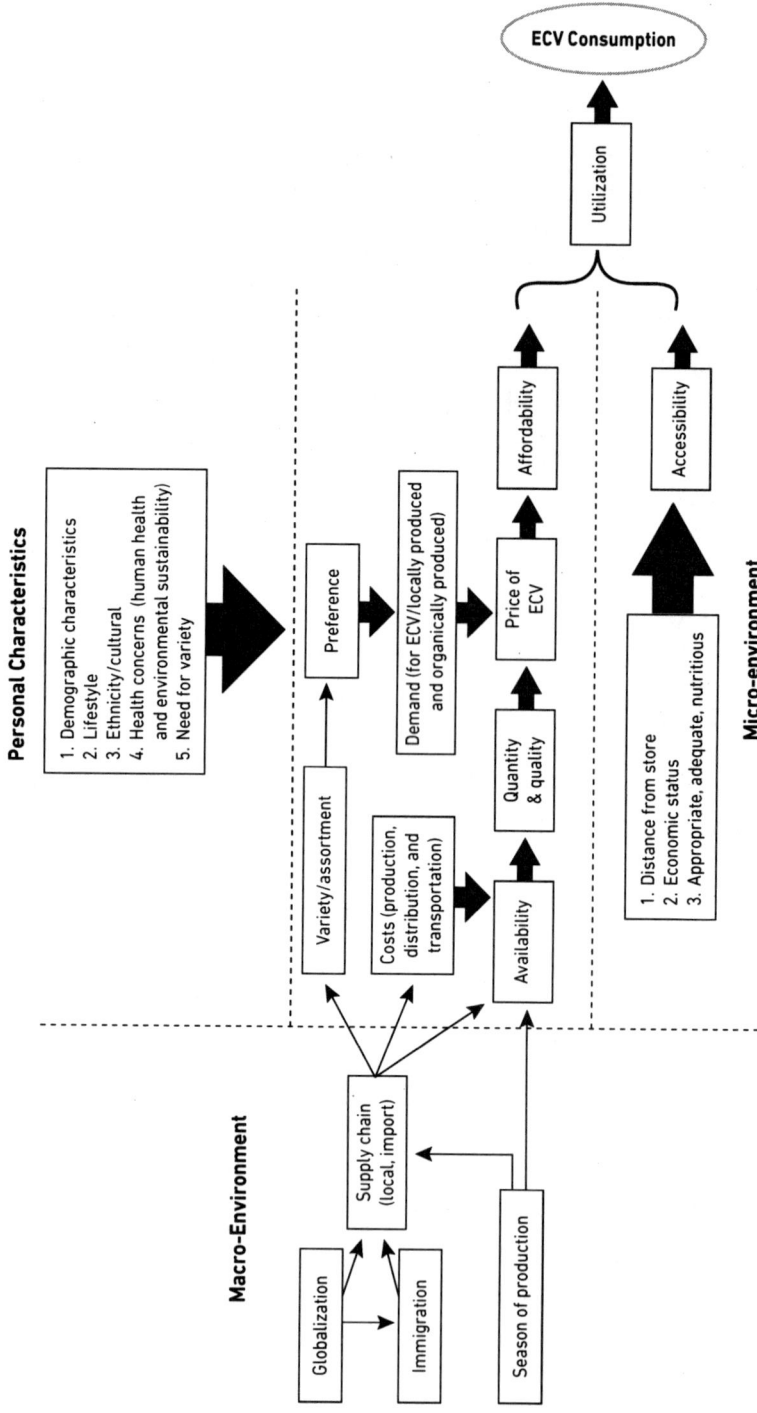

FIGURE 4.1 • Factors Affecting Ethnocultural Vegetable Consumption in the GTA

The Ethnocultural Vegetable Retail Market Structure

There has been an exponential expansion and heightened demand for ethnocultural vegetables in those parts of Canada where new immigrants have settled. This is expected to grow at around 12 percent a year in the next five years compared to other kinds of food (Hamilton and Spence 2008; Toronto Food Business Incubator 2011).

Local ethnocultural vegetable supply has been limited in the kinds of vegetables produced as some cannot yet be grown or may never be grown competitively in Canada. Locally produced ethnocultural vegetables are usually marketed through alternative markets such as the farmers' markets, farm shops, community-supported agriculture, and retail specialty stores. Local producers may also supply large chain stores, but their supplies are inconsistent due to low production during off-seasons, which is a disadvantage as the chain stores require a constant, year-round supply. To meet the annual demand, ethnocultural vegetables are imported in larger volumes from the United States, Mexico, and the Dominican Republic through large-scale dealers or the Ontario Food Terminal.

Vegetables are also imported in other ways such as through individual smuggling from Asia, Latin America, the Caribbean, and Africa. The size and frequency of this business is based on the demand and financial ability of the individual, who supplies ethnocultural vegetables directly to the dealers, such as specialty retail stores or restaurants. Since these supply chains are individually based, they have not been studied much and are not well understood. While this type of supply chain cannot be ignored, its sustainability and viability are questionable as the supply is in small quantities, costs are high, and profits low.

Although large-scale importation of vegetables has reduced costs and made vegetables more accessible, the continued dependency on importation and the regulations that govern the vegetable market have continued to favour more upscale retailers (Kajumba 2012). Consumer preference for supermarkets has put many smaller, ethnic stores out of business (Lister 2007). The preference for upscale supermarkets has also reduced the accessibility and affordability of these vegetables for some (Kajumba 2012).

Ethnocultural Vegetable Pricing

Pricing plays a major role in the affordability, accessibility, and consumption of ethnocultural vegetables (Powell, Zhao, and Wang 2009). Pricing is usually determined by consumer preference, proximity, seasonality, trade agreements (Kajumba 2012; Albrecht 2014), source of origin, packaging (Goodwin et al. 1988), production costs, consumer demands, and business costs (Waugh 1928). Consumers expect that price signals quality, but this is not necessarily the case in the ethnocultural vegetable market because of scarcity of and monopoly over certain crops. Consumers have continued to demand better quality, improved taste, food safety, hygiene, and improved size and shape (Waugh 1928), factors that have called for increased food prices.

Price setting involves consideration of the inputs and outputs so that the invested capital is reproduced with a profit, including for high-value crops like vegetables and organically produced crops, which are more expensive to grow. Large-scale producers' costs are minimized due to the scale at which they operate compared to small-scale producers' high production costs (Peterson et al. 1999). In small-scale retail marketing, operating costs, which are sensitive to quantity, are likely to be higher due to the small scale of production. Because of large-scale producers' minimized production costs, they can set relatively low prices and still meet their profit margins. Small-scale producers do not have this advantage. Their production costs are higher, yet their prices must remain competitive with the prices set by the large-scale producers and importers. In organic production, producers have managed to charge premium prices that cover their relatively higher costs of production.

The difference in vegetable prices creates two categories of consumers determined by their elasticity of demand. A product is price elastic if the responsiveness to a price change in terms of quantity demanded is more than the change. Consumers with a lower income tend to seek substitutes for some of the more expensive food or target sales when prices are relatively lower so they often compromise food quality for a cheaper cost. Higher income consumers, on the other hand, are less responsive to price and are more likely to pay premium prices for quality vegetables. Price is thus likely to play a major role among the low-income, high-price responsive customers, while quality may be of more importance among the wealthier high-income earners (Han and Wahl 1998). Furthermore, the place of purchase, whether super-market, discount store, or grocery store, may also affect the prices of vegetables.

The retail market for ethnocultural vegetables is diverse with different stores registering differences in variety, quality, and quantity of vegetables sold and thus the prices charged. The stores that were visited and discussed here (Kajumba 2012) were found to influence the vegetables accessed and hence the prices charged. In Toronto, large differences were recorded between prices of ethnocultural vegetables sold in different stores. Many convenience stores, located in the poorer, more marginalized neighbourhoods, are noted for overpricing their vegetables compared to the lower priced mainstream stores with more nutritious food in upscale neighbourhoods (TFPC 1996). The convenience stores in the low-income zones are usually stocked with more processed food, so consumers' ability to access basic, high-quality vegetables is very limited. This leads to more obesity and an increase in food-related illnesses, mostly affecting the young and the poor (Jackson et al. 2009). Dunn et al. (2011) also observed a price difference across stores depending on the class structure of the community. Their study indicated that prices are higher in high-income neighbourhoods as compared with low-income communities. With low availability of ethnocultural vegetables, those in poorer communities replace ethnocultural vegetables with less healthy, low-cost, more available processed food. How ethnocultural vegetables are displayed, packaged, and labelled, as well as their quality, assortment, and services provided also vary enormously from store to store (Kajumba 2012).

Retail Structure

The ethnic vegetable retail market continues to evolve with many small-scale traditional retailers now in the market. These stores often maintain displays that are not enticing or convenient. However, from observation and interview findings, the Chinese stores are better organized and have established food retail systems. In order to improve their performance, retailers have introduced displays that facilitate quick self-service. Mainstream stores have put this into practice, with Chinese stores taking the lead in establishing excellent displays. Mainstream stores and, to a greater extent, the Chinese ethnic stores have made an effort to design appealing fresh produce departments. Less effort has been made by most Afro-Caribbean stores, while South Asian stores are often between the appealing Chinese stores and the least appealing Afro-Caribbean stores (Kajumba 2012).

Compared with the major supermarkets, Afro-Caribbean and South Asian stores usually occupy a much smaller space and present a relatively poor product display. Afro-Caribbean stores are often the least organized and are therefore less able to facilitate a self-service system for customers (Kajumba 2012).

Packaging, Labelling, and Storage Facilities

The storage, packaging, and preservation methods of ethno-cultural vegetables in most ethnic stores need to be improved. Packaging is still very limited and cannot compete well with international standards and pressures nor with the arrangement usually available in supermarket foodscapes. The lack of freezers and cold-storage rooms means perishable ethnocultural vegetables that require cool storage are left standing on warm shelves, which shortens their shelf life and cuts into the store owners' profits.

Product labelling is an important component of the retail market, especially in a world of food scares where product labelling assures customers of quality and food safety. Consumers are not only interested in produce appearance and taste, but also the details of food origin and method of production through labelling as this tells much about how the environment is being preserved, where the produce was produced, and how the commodity's production may have impacted the producers of that commodity. Produce labelling is an important attribute in attracting consumers (Lumpkin, Greenberg, and Goldstucker 1985). Unfortunately, there are fewer properly labelled products in ethnic stores compared with the mainstream supermarkets. In Afro-Caribbean and South Asian stores, labels on most ethnocultural vegetables were usually poorly constructed. The labels often do not indicate the price or place of origin for many of the products. Different stores often label the same vegetable with different names. Labels that indicate place of origin are important for customers who are interested in purchasing local vegetables, but they also provide some assurance for food safety (Kajumba 2012).

Chinese stores are better at labelling their produce than most other ethnic stores, although not all the produce have labels that indicated produce name, place of origin, and price in Canadian dollars either per pound, kilo, or pack. However, these labels are rarely, if ever, on the individual package. Labelling in some Chinese stores is usually only in Mandarin or Cantonese. This may dampen the enthusiasm of customers who cannot read the labels. Most of the products in ethnic stores are sold fresh with minimal product differentiation in the ethnocultural vegetable market, while mainstream stores use product differentiation to facilitate choice for busy customers. Although stringent legislation is in place for food quality, a number of stores visited made no effort to maintain fresh, clean vegetables, an observation made across all the stores, particularly Afro-Caribbean stores, which had the highest numbers of poor-quality vegetables. In addition to fresh and packed food, ethnic stores are noted for diversifying to include health and beauty products, although these were still less diversified compared to mainstream stores.

Ethnic stores also tend to have less flexibility, opening for a maximum of nine hours with some opening for even shorter periods, while mainstream stores operate from 7 a.m. till 10 p.m., with some open for twenty-four hours for the convenience of busy customers. Nevertheless, ethnic stores remained superior in ethnocultural vegetable assortment and available varieties. Although the vegetable assortment in many mainstream stores has become diversified to include ethnocultural vegetables, the numbers of these vegetables sold are still limited with no guarantee that customers can find even the most preferred vegetables.

Clearly food systems in Canada have undergone changes as a result of globalization, immigration, and free trade. These have led to changing lifestyles and trading systems that have affected the availability, diversity, affordability, and utilization of ethnocultural vegetables by influencing pricing, supply chains, and retail outlets. The availability and diversity of the produce have increased, but affordability and accessibility have been restricted, as have their use and consumption. More retailers increase competition, which affects pricing. More widespread trade has increased consumer demands and expectations, leading to increased pressure on the retail market to continually improvise to remain competitive. This has left small retailers and producers under intense pressure to meet these high standards.

There are clear signs that retail agro-businesses are increasingly involved with many aspects of ethnocultural vegetable production both abroad and locally. Until quite recently, the produce was available only in ethnic stores, but they are now widely available at Loblaw, Sobeys, and Metro, and the quality competes well with even the best Chinese stores. Small and medium-size ethnic stores must struggle to avoid being wiped out or taken over. Loblaw's takeover of the high-quality Chinese T&T stores is a prominent example of this.

CHAPTER 5

Are Ontario's Farmers' Markets Sufficiently Inclusive?

Frances Dietrich-O'Connor, Bamidele Adekunle, and Glen C. Filson

Most of the vegetables available at southwestern Ontario markets are fresh and wholesome. In this chapter we assess the accessibility of this food. We also describe the degree of interest in ethnocultural vegetables expressed by the mainly white residents of Guelph, a city of about 150,000, an hour outside the Greater Toronto Area (GTA).

Short supply-chain farmers' markets have been proliferating to counter problems such as limited amounts of fresh local produce and the disconnect between producers and consumers of food in the globalized food regime. Farmers' markets provide an "alternative system of food production and distribution that is not based exclusively on the commodity relationship and profit maximization" (Kirwan 2004, 398).

Ethnocultural vegetables are mostly imported from abroad, though more are now being grown in southern Ontario and the extent of this market is being recognized. One way of reaching consumers directly would be for farmers to sell their locally produced ethnocultural vegetables at farmers' markets. Unfortunately, there are few outlets in Ontario other than the Ontario Food Terminal

Farmers' Market and the St. Lawrence Market in the GTA. (The McVean Farmers' Market was active during our research, but is now defunct.)

There is a significant cultural barrier between horticulturalists in Ontario, who are mostly of European descent, and the main consumers of ethnocultural vegetables, many of whom are more recent immigrants. This is beginning to change as exposure to these foreign cultures is creating a change in some farmers' mindscapes[1] and the landscape of farming practices in Ontario. New Canadians can get these locally grown, fresh ethnocultural vegetables at only a few farmers' markets. Apart from providing fresh and organic fruits and vegetables, a farmers' market is where consumers can interact with and advise farmers about the attributes they want in the produce they purchase.

We wondered whether new Canadians are as willing to purchase food from outlets other than farmers' markets. In California, Alkon and McCullen's (2011) study of a farmers' market found that the market promoted *white* practices and discourses. Campigotto's 2010 study of Toronto's Wychwood Artscape Barns Farmers' Market also found that while the produce was fresh and relatively expensive, its clients were mostly well-off whites in contrast to the surrounding ethnic and economic diversity of the neighbourhood. If a greater diversity of world crops were available, this would likely increase the diversity of the clients of farmers' markets.

Alkon and McCullen (2011) conducted an ethnographic study at two farmers' markets in northern California. Their study sought to understand how whiteness is both performed and perpetuated at farmers' markets. To start, Alkon and McCullen (2011, 950) identified the "romantic imagery surrounding small farmers as well as the imperative to buy directly from them." The article challenged this imagery, asserting that it ignores the historical role of race in American agriculture and "leads us to believe that the whites we see selling at the farmers' market, rather than their mostly Latino/a employees, are those who presently grow our food. Farmers' markets such as those we study emphasize the importance of building community, but are often unaware that they define community in a way that draws in whites while pushing away people of color" (Alkon and McCullen 2011, 950). Alkon and McCullen (2011, 950) perceive alternative food movements perpetuating whiteness through discourses that "paint alternative food choice as a moral rather than economic decision and normalizes affluence."

In another study of farmers' markets and community shared agriculture managers, Guthman (2008b) identified similar ways in which discourses of alternative agricultural movements may be responsible for the prevalent whiteness of such movements. Guthman (2008b) found that the language used by several managers interviewed provided important examples of two manifestations of whiteness. The first is the idea that "color blindness or the absences of racial identifiers in language are seen as nonracist" (Guthman 2008b, 390). As Guthman (2008b, 391) asserts, this colour blindness "does its own violence by erasing the violence that the social construct of race has wrought in the form of racism." The second manifestation of whiteness identified by the study is universalism. This universalism is in the assumption that those values held predominantly by white people are normal and possessed by most other people (Guthman 2008b).

In a second article on the topic, Guthman (2008a, 431) examined a number of students interested in taking part in the alternative food movement and "bringing good food to others." In this study, Guthman (2008a, 431) found a number of similar themes. This included students' assertions about the universalism of the alternative food movement and a discourse that "reflects whitened cultural histories."

Few Ontario horticultural farmers are aware of the economic benefits of cultivating these specialized crops for new cultural groups (Adekunle, Filson, and Sethuratnam 2012). One main challenge is that new Canadians might not be able to afford to purchase from farmers' markets. Moreover, most people from the Afro-Caribbean, Chinese, and South Asian groups interviewed by ECVOntario were interested in purchasing their culturally preferred vegetables so long as they could obtain them cheaply. A recent study by the University of Guelph's Food Institute and Dalhousie University interviewed over one thousand customers in May 2016 only to find that recent price increases in fruits and vegetables have meant that younger people with relatively less education and lower incomes had reduced their purchases of produce over the past six months (Charlebois 2016). While their study did not identify people's ethnicity, it is clear that many people with these characteristics—and certainly recent immigrants tend to be relatively poorer than average—do not perceive farmers' markets as the most affordable place to obtain their cultural food.

In order to understand the availability of ethnocultural vegetables at farmers' markets, we posed the question: What determines what is sold at farmers' markets? Although some external factors such as climate or rules imposed by farmers' markets play partial roles in shaping what is sold at these markets, generally it is assumed that the types of vegetables sold are determined by farmers' crop choices. Although extensive literature exists to explain farmers' crop choices, currently there is little or no literature to explain how farmers choose which products to sell at farmers' markets. So, we turned to the literature on farmers' crop choices in order to understand and explain the relative lack of ethnocultural vegetable availability at farmers' markets in Ontario.

Farmers' Crop Choices and Decisions about What to Sell at Their Farmers' Markets

Historically, literature on farmers' crop choices has been dominated by neoclassical economic approaches: farmers act as autonomous, rational, free agents able to choose between several actions.

These rational actors will make choices based upon their knowledge of the market and use this knowledge to maximize their profits (Darnhofer, Gretzmacher, and Schneeberger 2006).

Ravnborg and Rubiano (2001) also assert that farmers' crop decisions are largely driven by profits. In order to make the desired profits, the ease of selling crops plays an important role in farmers' decision making and they are particularly interested in crops for which they have a secure buyer. A number of economic factors play important roles in determining farmers' decisions. These factors include crop sale price and income relative to efforts, crop profiles with desirable economic attributes, ease of transportation, low-input requirement, labour requirement, free time and allowance (Briggs 1985; Raynborg and Rubiano 2001; Van Huylenbroeck and Damasco-Tagarino 1998).

Considering the nature of farmers' markets, farmers' decisions are likely influenced by demand factors. Farmers and consumers alike appreciate that farmers' markets provide an opportunity for interactions between farmers and consumers. However, there is little research that attempts to understand how this particular relationship influences farmers' decision making.

In a study of the experiences and perspectives of farmers from upstate New York farmers' markets, Griffin and Frongillo (2003) found that farmers liked selling at farmers' markets because it gave them the opportunity to try out new products or crops on a small scale. As Hunt (2006, 63) notes, "Compared to agricultural systems based on long supply chains, farmers' markets farmers have the opportunity to interact directly with consumers and respond to shifts in demand as quickly as the agricultural cycle will allow." Feagan, Morris, and Krug (2004) point out that community shared agriculture, farmers' markets, and food box programs bring producers and consumers closer together physically and socio-culturally.

Much of the interest in farmers' markets and local produce has revolved around the concept of *embeddedness*. According to

Feagan and Morris (2009, 237), "*Spatial embeddedness* includes a group of motivations associated with the desire to buy food produced locally, and where the producer and consumer are more directly linked. Freshness fits in this category." Such programs are also created in part to support local farmers and rural communities. In their study of farmers' markets in Brantford, Ontario, Feagan and Morris (2009) found that price was not an important motivation for purchasing for three-quarters of the consumers they surveyed, probably because those buying food at farmers' markets are either relatively well off or they like the ambience.

While the *Local Food Systems*' authors (Martinez et al. 2010) acknowledge that farmers' markets generally provide healthier food because it is fresher, they also affirm that there is no solid empirical evidence for this.

In a study of fourteen farmers' markets in Toronto, Bond and Feagan (2013) determined roughly how much culturally appropriate food was available. Managers of these farmers' markets told them that "farmers are generally risk-averse and therefore cautious regarding the introduction of new products like culturally appropriate foods" (Bond and Feagan 2013, 7). Curiously, the vendors at these markets are often migrant workers from tropical countries where there is an abundance of the ethnocultural vegetables that are rarely seen in the farmers' markets of the GTA with the exception of the OFT Farmers' Market and the St. Lawrence Market. This is despite the more diverse consumers in the GTA. Instead these ethnic consumers usually have better luck obtaining their culturally appropriate food at their local ethnic grocery stores (Bond and Feagan 2013).

Desrochers and Shimizu (2012) caution that farmers' markets sometimes have people fraudulently reselling products that are not local, which can result in major fines for farmers. One Guelph Farmers' Market vendor told us that while he produces all the food he sells at the market, he believes that most of the food some vendors sell at the Guelph Farmers' Market originates from the Ontario Food Terminal. Though we were unable to verify his claim, some farmers' markets may accept this reselling practice, but most probably do not. It is particularly egregious if the vendors acquire the produce at wholesale prices from abroad, then resell them at high farmers' market prices during the summer months when local produce is available.

Farmers selling at farmers' markets are influenced by their direct interaction with consumers (Hunt 2006). One farmer indicated that he keeps detailed information on demand for crops from his community shared agricultural operation and farmers' market sales, and uses that information to decide which crops to grow. The demand side factors are very important in vendors/ farmers' choice of produce to sell at the farmers' markets. Ruel, Minot, and Smith's (2005) conceptual framework provides an important starting point for understanding the demand factors influencing the supply of vegetables at farmers' markets. They argue that the main factors determining demand are household incomes, the price and availability of fruits and vegetables, and household members' preferences. For Asfaw (2008), household incomes and prices are also main considerations, but he asserts that non-economic factors such as the media and food advertisements can also play important roles in determining the demand for vegetables. Figure 5.1 provides a graphical depiction of the various factors determining the availability of ethnocultural vegetables at farmers' markets.

Conceptual Framework

This framework suggests that apart from the economic explanations for demand and supply that determine which ethnocultural vegetables will be available at the market at any given point, there are other factors impacting the purchase and sale of this produce.

The prevailing wage rate in the country, the gross domestic product (GDP), and education will affect consumers' income. Also, consumers with a high level of education also tend to have a better income if the economy is productive and there is no natural or managerial disaster.

Some high-income consumers may still not patronize farmers' markets because the ethnocultural vegetables they want are not available there. The prices of some ethnocultural vegetables

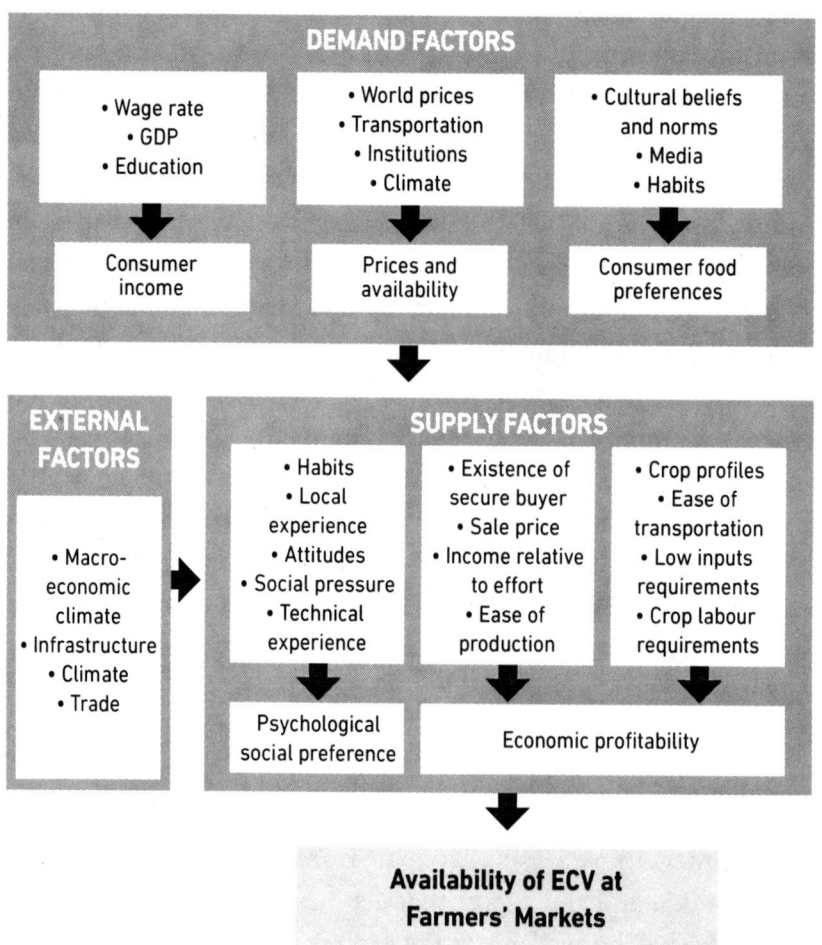

FIGURE 5.1 • Determinants of Availability of ECV at Farmers' Markets

can also be affected by world prices, especially if they are imported from other countries where these crops are readily available and are produced in large quantities. Major expenses incurred as produce moves through the international global chain include agricultural extension services, assembly, transportation, and other costs. Appropriate policies might make the local version of a vegetable more affordable and readily available in the farmers' markets by eliminating the transportation cost of the imported version while also improving freshness and quality.

Nicaraguan okra sells for upwards of $25 a packet at the Ontario Food Terminal, but it has been flown to Toronto from an American-owned Nicaraguan farm. Local okra has, however, rarely met the quality or at least the presentation of the handpicked and sorted Nicaraguan okra, so its price is often lower. Once the local version's picking, sorting, and packaging can compete, okra will sell for less and greenhouse gases will be reduced (see Aitken 2014).

Varying weather conditions in Canada and abroad, coupled with the fact that not all ethnocultural vegetables can be successfully grown in Canada, negatively affect price and availability. Another important factor that shapes demand is consumer food preferences.

Consumer food preferences are usually affected by cultural beliefs and norms, which at times affect how they acclimatize to the food system in their new country (Adekunle et al. 2011). Some people will always prefer to patronize ethnic stores because of kinship and trust. They enjoy visiting these stores as a social experience as well as to purchase food. The media can also transform consumers' perceptions. For example, the GTA Chinese, South Asian, and Afro-Caribbean consumers interviewed for Chapter 2 usually indicated that the types of advertisement, especially flyers, affect their shopping decisions. Consumers' habits might also affect their preferences and contribute to economic development when culturally appropriate food is available in the market (Adekunle, Filson, and Sethuratnam 2012). For example, consumers from different cultural groups might think that farmers' markets are for the elite and once this illusion is reinforced, it becomes a reality that a farmers' market is expensive and does not have the ethnocultural vegetables they want. In Figure 5.1, the factors that affect demand for ethnocultural vegetables in farmers' markets are consumer income, prices, availability, and consumer food preferences. To complete this framework, we need to explain the factors that decide whether these ethnocultural vegetables will be supplied.

The ethnocultural vegetable market is "demand pulled." In other words, demand precedes supply once factors such as psychological/social preferences and economic profitability are favourable. Psychologically, farmers' habits and local experience will determine if they grow a new crop and market it at the farmers' market. Their decisions are also affected by their peers, who may pressure their colleagues to explore new opportunities, which affects farmers' and vendors' attitudes toward and technical experience with the produce. Another major supply factor is economic profitability. Farmers/vendors at a farmers' market will sell a vegetable only if it guarantees profit. Farmers will produce or market ethnocultural vegetables only when they are confident there will be buyers and that they can sell the produce at reasonable prices, even though consumers from some cultural groups might think those prices are high. In addition to profit, the food must be relatively easy to produce. Ease of production improves as the number of years spent on cultivation increases. Ease of transportation will also increase profitability and creates incentives for local farmers and vendors to supply ethnocultural vegetables at farmers' markets.

Other than the demand and supply factors, other factors such as macro-economic conditions, infrastructure, climate, and trade affect farmers' markets' exclusion of imported produce. Macro-economic variables usually have a trickle-down or multiplier effect on most activities in any economy. Trade affects the supply of ethnocultural vegetables in the Canadian food market, but it is not likely to play a significant role in farmers' markets if they continue their policy of selling local products. The provision of public goods and infrastructural support may create an enabling environment for the availability of ethnocultural vegetables. Climate is beyond the control of all the stakeholders in the market. The drought situation in 2012 and 2016 that affected much of North American agriculture increased the price of locally produced ethnocultural vegetables.

Farmers' Market Research Methods Used for This Study

A combination of qualitative and quantitative methods was employed to examine the availability of ethnocultural vegetables at farmers' markets and the factors influencing their availability or scarcity. This chapter draws on direct observation at the Guelph Farmers' Market and the Dufferin Grove Organic Market in Toronto, market data on the availability and pricing of ethnocultural vegetables at these venues, as well as participant observation at the Guelph Farmers' Market from May 2011 to July 2012, followed by interviews with five of its vendors. One-time direct observations were also completed at Sauraren Market in Toronto, The Stop's Green Barn Farmers' Market, the Elora Farmers' Market, Toronto's St. Lawrence Market, and the North York Farmers' Market. The St. Lawrence Market had the best supply of ethnocultural vegetables.

The downtown Guelph Farmers' Market is open on Saturdays, and has a wide variety of vendors, artisans, and producers. Unlike other markets, it does not require that vendors sell local products exclusively, which accounts for the abundance of Ontario Food

Terminal produce. By contrast, the OFT's Farmers' Market fines vendors for selling imported produce.

The Dufferin Grove Organic Market, located in Toronto's west end, also runs a weekly market. The organic produce can be local or imported. Direct observations were completed on four occasions at each market between May and August 2011, while ethnographic research has continued in Guelph up to the present.

Direct observations were made of market patronage and the general comportment of customers and vendors. In addition, the researchers engaged in informal conversations with vendors to gain an understanding of their crop decisions. During these interactions, farmers were asked how they decided what crop varieties to sell. Vendors were also asked if they have grown ethnocultural vegetables in the past and if they would consider growing them in the future. Those selling them were asked why they chose to do so and if they faced any challenges in growing the crops. Interviews with busy vendors were brief and informal, though longer interviews were later conducted with five more Guelph Farmers' Market vendors about the potential for customers of European descent to purchase ethnocultural vegetables. In addition to analyzing 365 interviews with Guelph residents at Stone Road Mall, Davis (2015) also conducted two focus groups with thirteen University of Guelph undergraduate and graduate students.

During the visits to the markets, records were also kept on the availability of twenty-four high-demand ethnocultural vegetables. This list was based on high-demand vegetables identified in *Growing International: Exploring the Demand for Culturally Appropriate Foods* (Filson et al. 2011). Using this data, the current study gathered information on the top ten vegetables identified by each ethnic group. Due to preference overlap among the ethnic groups, only twenty-four ethnocultural vegetables were identified for the study.

This list consists of vegetables that are not traditionally part of a Canadian diet, such as cassava and bitter melon, in addition to some commonly found in most Canadian diets (such as tomato and potato). The study also gathered information about vegetables identified by each group as "often lacking in availability." In addition to availability, data were also collected wherever possible on prices, origins, and organic or non-organic status of the vegetables.

Findings

We faced a number of challenges in the field. It was difficult to gather reliable and comparable price data from the farmers' markets because each farmer used a different unit of measurement for prices. Some farmers priced items by the pound while others priced them by the kilogram, pint, or quart. These measurement issues presented challenges for data analysis, as some data could not be compared across time or marketplace. In addition, at times prices varied significantly among vendors in the same market.

The price data were compiled and analyzed using descriptive statistics to develop a preliminary understanding of crops' price fluctuations over the period of study. Figure 5.2 shows the price fluctuations for a number of vegetables studied. The prices do not seem to follow a clear trend, with some vegetables experiencing considerable spikes in prices (celery, for example) while the prices for others (such as potatoes) remained relatively stable throughout the study period. While price fluctuations may be caused by numerous factors, the seasonality of particular crops likely played the most important role.

Most of the produce available at the markets were mainstream vegetables such as tomatoes, onions, and potatoes, which are already part of the Canadian diet as their source is mainly North American or European. As Table 5.1 indicates, fifteen of

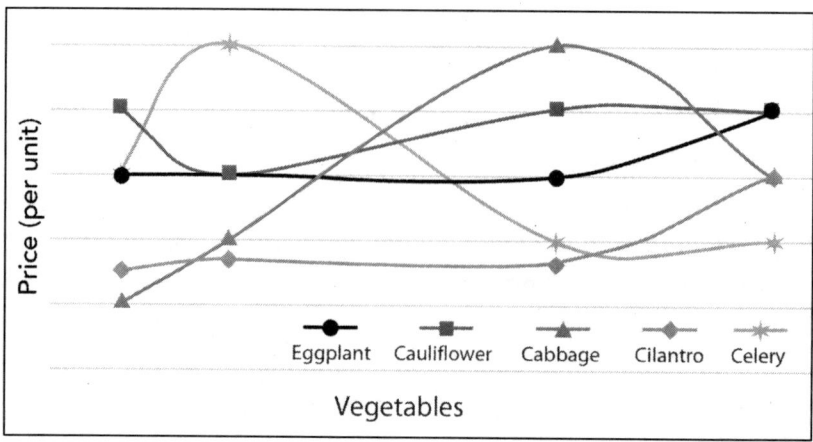

FIGURE 5.2 • Prices of Vegetables at the Guelph Farmers' Market per Unit
(May–August 2011)

the thirty-one vegetables for which price data were supposed to be collected were not available during any of our farmers' market visits. Perhaps not surprisingly, the vegetables not found during the market visits were less mainstream vegetables.

Table 5.2 indicates the availability in percentages (based on the number of visits) for vegetables that were available during visits. Of the vegetables for which price data were collected, the most commonly available were potatoes and onions, which were found at all the farmers' markets. Even bok choy and eggplant, which are produced in Ontario much more than other ethnocultural vegetables, are not always available at farmers' markets.

TABLE 5.1 • Vegetables Not Available at Farmers' Markets

Okra	Plantain	Fuzzy melon
Chinese broccoli (*gai lan*)	Cocoyam leaves	Drumstick
African eggplant	Yard-long beans	Indian squash
Smooth amaranth	Cassava	Pointed gourd
Bitter melon	Bamboo shoots	Bitter leaf

Not all the produce at the Guelph Farmers' Market is locally produced because some, such as plantain, cannot be grown in Canada. To support the local economy, the vendors at farmers' markets try as much as possible to sell locally produced food. Most transactions in the market are cash based, though a few vendors accept debit cards.

The farmers' market is a meeting place where people can interact. The sense of community at the farmers' market cannot be overemphasized. Those who shop at the market appear to be relatively well off, health conscious, and include many University of Guelph faculty, middle or senior administrative staff members, as well as some students. While most of the consumers appear to be at least middle class (professionals, managers, and salaried

TABLE 5.2 • Vegetable Availability

Rank	Vegetable	Percentage of Markets
1*	Potatoes	100
1*	Onion	100
2	Tomatoes	87.5
3*	Bok choy	75
3*	Carrots	75
3*	Cabbage	75
3*	Sweet potato	75
3*	Snow peas	75
3*	Green beans	75
3*	Pumpkin/squash	75
4*	Eggplant	62.5
4*	Cauliflower	62.5
4*	Cilantro	62.5
4*	Celery	62.5
5	Spinach	50
6	Chinese cabbage	12.5

Note: *Represents a tie.

working class), this is based on participant observation for more than a year and not hard evidence from participants at the Guelph Farmers' Market. Most of the very few people from the African, Chinese, and South Asia cultural groups who frequent this market are affiliated with the University of Guelph.

The farmers' market will contribute to economic development and promote local food production if it becomes more inclusive with respect to ethnocultural vegetables as well as relatively more recent immigrants. The sense of community is encouraging and if the market becomes more inclusive, it might make acculturation of different cultural groups to the mainstream culture easier to achieve. The market is usually full of activities such as entertainment by musicians, and is also a meeting place and a location for information dissemination. It is a place where consumers can learn about how healthy food can mitigate or help prevent diabetes, cancer, and cardiovascular diseases.

Quality is maintained in the Guelph Farmers' Market and food safety is a high priority. This is the standard because sellers are concerned about their goodwill and reputation, and because buyers will make repeat purchases only if quality and food safety are not compromised. To increase the advantages of these specialized markets, the government needs to make the public more aware of their benefits, backed up with an institutional framework that supports economic growth through the consumption of locally produced vegetables. Availability fluctuates and products from different farmers might not necessarily be of the same taste and value. The farmers' market is usually busy during the harvest period as a result of the increase in available crops and the nice weather. It is a very good period for the locavores because some of the produce are harvested the same day as they are sold at the market. It is pertinent to note that discussion about freshness became pronounced during this period.

Farmers try to create awareness by educating buyers on the benefits of their products. This often establishes trust and leads buyers to purchase from the same farmers. After three months in the field, we developed relationships with different vendors in the market. The relationships became so strong that the vendors often knew which produce and what amount we wanted to purchase.

Bok choy was the only somewhat non-traditional vegetable commonly found at the farmers' markets. It was more available at the Dufferin Market and in fact was found during each observation. In Guelph, on the other hand, bok choy was available only during half of the observations. In addition to bok choy, Chinese cabbage was found a few times. Interestingly, snow peas, which Filson et al. (2011) identified as "often lacking in availability," were relatively available at the farmers' market.

Farmers who were selling less commonly available vegetables such as bok choy, Chinese cabbage, and snow peas cited a number of reasons for choosing to do so. One farmer chose to grow bok choy for its health benefits. Another farmer who sold bok choy and Chinese cabbage explained that the farm he worked for had been growing the crops since he began working there. He said he liked to grow a wide variety of different crops. In addition, he noted that of all the crops he grew, bok choy and Chinese cabbage were his favourite to grow because "you can put them in the ground and forget them" (personal communication, 2011). He explained that because he really liked the crops, he tended to push them to the front of his booth to encourage consumers to purchase them.

Crossover Impact of Ethnocultural Vegetables in Guelph

To corroborate our suspicion that there are more ethnocultural vegetables available elsewhere, thus increasing the potential cross-ethnic impact, in the winter of 2012/2013, we interviewed

365 respondents at Guelph's Stone Road Mall in a survey about their perception, identification, and adoption of these vegetables. Guelph's population is vastly different from the GTA, with far fewer South Asians, Chinese, and Afro-Caribbean people, but because we work at the University of Guelph, we thought it would be important to determine how aware Guelph's European-descent population is regarding ethnocultural vegetables. We assumed that if we could see a significant crossover effect within Guelph, the effect would surely be even higher within the GTA. Of the 365 respondents, 73 percent were born in Canada, 3 percent in China, 2 percent in the United Kingdom, 2 percent in India, 1.4 percent in the United States, 1.1 percent in Italy, 1.1 percent in Philippines and a few other countries.

Research conducted by Keisha Davis (2015) on the potential crossover demand for ECV in Guelph among Canadians born in Canada found that the relative lack of availability and quality of ethnocultural vegetables in Guelph often limited the extent to which Canadians of European descent were willing to purchase them. Eggplant, bok choy, and okra were the most widely recognized and most readily tried by non-immigrant Guelph attendees at the Guelph Stone Road Mall. Having sufficient knowledge about what ECV are and how one must cook them, combined with freshness and being locally produced, enhanced the crossover impact of these vegetables. Davis, a Jamaican citizen, assumes that the crossover effects of ECV will continue to grow within areas like Guelph.

A recent visit to the No Frills store on Silvercreek Parkway and the Metro store on Stone Road, both in Guelph, provided us with empirical evidence on the mainstream stores' recognition of culturally appropriate food. The No Frills store now has a section for vegetables specifically for people from South India and Sri Lanka, while a few years ago Metro created a section for cultural food popular among the Chinese, South Asians, and Afro-Caribbean. As supermarkets carry increasing amounts of newly demanded ethnic food, all customers are encouraged to try them (Wright, Nancarrow, and Kwok 2001). This means that individuals not included in a particular ethnic group are able to access non-traditional ethnic food. Since many kinds of new ethnic food were viewed positively by individuals outside those ethnic groups (Bäckström, Pirttilä-Backman, and Tuorila 2003), there has obviously been a crossover effect where individuals from the host country have adopted some ethnic food from immigrant communities.

A crossover survey of 2012/2013 of randomly selected people in Guelph's Stone Road Mall was designed to ascertain the cross-cultural impact of ethnocultural vegetables. The average age of the 365 respondents, mostly women, was thirty-eight, and the average number of years spent in Canada was thirty-two. They are literate, mostly non-vegetarian, and most households had more than three residents. Respondents spent an average of $45 per month on ethnocultural vegetables, and 16 percent of the respondents were willing to spend even more on these crops.

Over 94 percent of the respondents preferred to purchase their vegetables in supermarkets, 54 percent like to go to the local farmer's market, and 27.7 percent grow some vegetables in their farms or gardens. Only 3.9 percent access Community Shared Agriculture programs to get vegetables (see Chapter 6). The ethnic stores are not very popular with most Canadian-born people in Guelph. Only 14.4 percent of respondents were willing to buy vegetables there. Our suspicion is that cultural stores are usually visible in large and very diverse cities such as Toronto, Mississauga, Montreal, and Vancouver. The source of purchase is not as important as why people will purchase a particular product. Most of the respondents consume vegetables for their health benefits, pleasure, ethnicity, taste, and availability. In as much as the food is good, respondents want the product to be of excellent quality. Figure 5.3 depicts the quality improvement desired by the respondents.

Thirty-five percent of the respondents expect improvement in freshness in the vegetables in their area. Twenty-one percent preferred to buy local vegetables, and 17 percent of the respondents would prefer it if organic vegetables are readily available. Fourteen percent thought the quality of vegetables needs to be improved, and 13 percent of the participants found it difficult to access their desired vegetables. Of those respondents who love vegetables, we wanted to know if they would be willing to explore new ones. Indeed, they said they would. Around 90 percent of the respondents were willing to try new vegetables, a result of mingling with people outside their group. If the ethnocultural vegetables are locally produced, a higher percentage would be willing to try them.

However, Davis found that both availability and the quality of ECV are inadequate. Local Guelph folks' knowledge and awareness

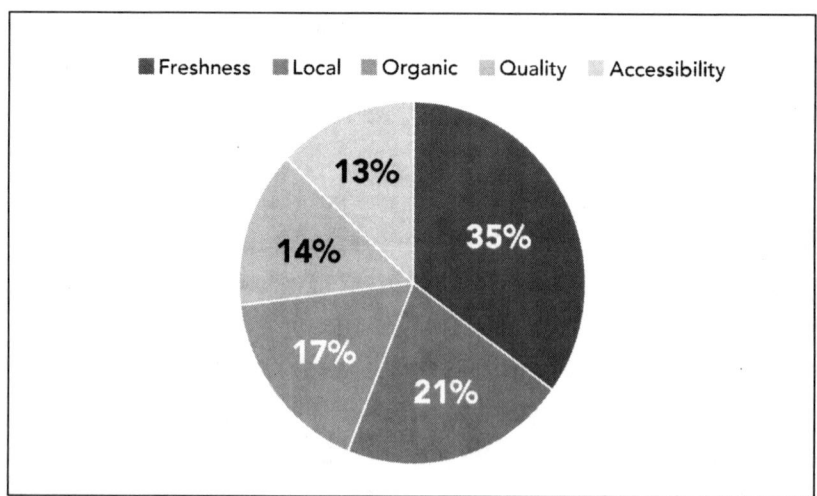

FIGURE 5.3 • Desired Vegetable Improvements

of ECV are also quite limited relative to those in the GTA. But whereas "Canadians were found to be adventurous and willing to try newly introduced ECV," the ethnic group respondents were not as willing to try other ethnic groups' vegetables (2015, 122).

To ascertain the cross-cultural impact of ethnocultural vegetables in multicultural Canada, we tested respondents' ability to identify the crops. The results are presented in Figure 5.4.

Based on our survey, Chinese eggplant was the most identified (78.9 percent) by the respondents, though it is presumed to be a Chinese ethnocultural vegetable. Bok choy and okra were identified by 50.4 percent and 39.2 percent, respectively. However, only 3.3 percent were able to name Chinese broccoli even though many people said they ate the vegetable but did not know its name. Most people did not know what cassava was, nor do they consume it.

Nevertheless, we believe that the cross-cultural impact can increase the demand for some of these vegetables. There is obviously a need for better market linkages as the study of the Guelph Farmers' Market reveals.

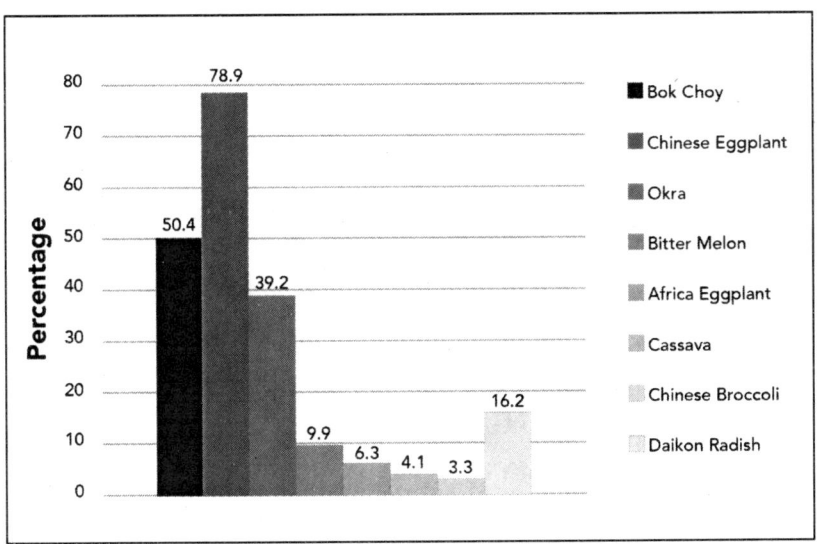

FIGURE 5.4 • Identification of Ethnocultural Vegetables

Discussion

As the study of the crossover effects at Guelph Stone Road Mall reveals, there is growing demand for ethnocultural vegetables outside of Toronto. Local ethnic stores like the Chinese-owned Taskgo's as well as the larger supermarkets are beginning to respond to this demand, yet the same cannot be said for the farmers' markets. The Mall study shows that we will experience greater cross-cultural impact of these vegetables when there is increased education, international travel, and interaction between the mainstream population and their multicultural neighbours. Based on our sample from Guelph, a good education, plus acculturation, curiosity, and other cultural or economic factors lead to a high level of cross-cultural impact.

We found that the availability of ethnocultural vegetables was very limited at the two farmers' markets we systematically visited, nor were they found during our occasional visits to other farmers' markets except for Toronto's St. Lawrence Market, McVean Farmers' Market, and the Ontario Food Terminal Farmers' Market. Some specialized farmers' markets such as FarmStart's McVean Farmers' Market in Brampton, Ontario, provide particular ethnocultural vegetables such as okra and eggplant. A few of the farmers sell most of the amaranth produced through direct contacts rather than through the farmers' markets. A specialized farmers' market may be a way to make it more inclusive. Farmers' markets specifically designed for special groups, such as the New Roots Farmers' Market in San Diego (Brown 2011) and FarmStart's market in Brampton are successful examples. These few initiatives notwithstanding, there are still serious challenges facing those who want to introduce ethnocultural vegetables into mainstream farmers' markets. The most available were ones that would traditionally be found in a Canadian diet, such as tomatoes, onions, and potatoes.

Further research would be required to understand the scarcity of ethnocultural vegetables at farmers' markets. As the literature review indicated, there are a number of factors that influence what vegetables are available at farmers' markets. These include consumer income, crop prices, consumer preferences, farmers' psychological or social preferences, and crop economic profitability. From our informal interviews with farmers at the markets, a few important themes emerged. To start, the lack of a reliable market seemed to be an important factor for a few farmers. For example, a farmer mentioned that he used to sell amaranth, but found that large-scale marketing is difficult, so making a profit at the prevailing market price for amaranth was not feasible.

This may be because there is very little funding for farmers who want to market their products locally whereas there are much larger federal and provincial subsidies for medium- and large-scale farmers who export their products. The farmers' safety net is based on a rolling average of how much the farmer has earned over the past several years compared to what he or she earns in the current year so that the net provides a floor below which the farmers' incomes do not fall.

Some farmers had previously grown ethnocultural vegetables, but a number of barriers had led them to scale back their production: lack of information about how to access markets, perceived lack of sufficient demand for these vegetables at farmers' markets, and perceived difficulty in breaking into the ethnocultural vegetable market. Some farmers felt that at current price points it was very difficult to make a profit by competing with imported ethnocultural vegetables.

The issue of whiteness in Canadian farmers' markets has been studied. In an unpublished master's thesis, Campigotto (2010) examined the demographic diversity of the Wychwood Artscape Barns Farmers' Market. Through her study, she found that although the market's neighbourhood is very ethnically and economically diverse, the market's patrons were largely university-educated, white, high-income individuals. This is also consistent with our ethnographic study of Dufferin Grove Organic and Guelph Farmers' Markets.

The aforementioned authors are careful to emphasize that they are not asserting that particular foods or alternative food practices are inherently white. Nonetheless, these studies do indicate that there are a number of practices that exclude many non-white individuals from participating or wanting to participate in farmers' markets as well as other alternative food movement activities. The current research has explored the factors that determine the availability of ethnocultural vegetables at farmers' markets. Further research is required to determine which factors have had the largest impact on the availability of this produce in farmers' markets. While not the only important factor, the growing literature on whiteness and farmers' markets suggests that there may be a relation between the relative whiteness of farmers' markets and the availability of ethnocultural vegetables.

Some authors point out that the local food system is flexible and capable of creating hybrids (Hall and Mogyorody 2002; Mount 2012). Motivating factors for participating in community shared agricultural (CSA) operations may suggest that there is flexibility in what consumers expect from an alternative food system. Some motivating factors, such as the demand for local, high-quality produce that is part of a larger environmentally friendly system of food production, are unlikely to change, resulting in shareholders' purchasing produce from mainstream stores. The CSA model appears to be flexible enough to adapt to the changing demand of consumers, their communities, and wider society.

CHAPTER 6

Community Shared Agriculture and Its Impacts on Culturally Appropriate Food Availability

Monika Korzun, Bamidele Adekunle, and Glen C. Filson

Community shared agriculture is making the agricultural landscape and mindscape of Ontario more conducive for local production and consumption of culturally appropriate food. We begin with a brief history of community shared agriculture (CSA), its structure, statistics that inform policy prescriptions, and the role of CSA as an alternative food regime.

History of Community Shared Agriculture

It originated in Japan in the 1960s as the *teikei* movement (Schnell 2007) when a group of women concerned with food quality and growing urban centres partnered with local farmers to protect the farmers, their land, and their health (Imhoff 1996; Sumner, Mair, and Nelson 2010). *Teikei* means "partnership" or "co-operation." These women approached local farmers and in exchange for financial capital, the farmers agreed to convert to organic farming (Schnell 2007; Brehm and Eisenhauer 2008;

Feagan and Henderson 2009; Shi et al. 2011). For the Japanese, *teikei* has since come to mean "food with the farmer's face on it" (Imhoff 1996).

In the 1960s, Germany and Switzerland operated under similar models. When the model reached North America in the 1980s, it was termed "community supported agriculture" by American organic farmer and food activist Robyn Van En (Schnell 2007; Shi et al. 2011). The first documented community shared agriculture in the United States was established by Robyn Van En, Jan Vander Tuin, and John Root Jr. in 1985 in the Berkshire Mountains in western Massachusetts (Cone and Myhre 2000; Schnell 2007). Now there are more than thirteen hundred registered CSA farms in North America (Robyn Van En Centre 2009, cited in Sumner, Mair, and Nelson 2010). In 2007, 12,549 farmers in the United States reported marketing their products via community shared agriculture (Community Supported Agriculture 2009). In 2002, there were about five hundred CSAs in Canada (McAdam 2002, cited in Haranandani 2009) and there are many more now.

The Structure of a Community Shared Agriculture

In community shared agriculture, consumers pay a set price at the beginning of the farming season for a weekly share of produce in return. In southern Ontario, the shares are distributed during the growing season, which tends to be from mid-June to mid-October (Ferris and Behmann 1994). The food may include vegetables, fruits, herbs, meats, eggs, and dairy products. Fish, grains, nuts, honey, jams, other preservatives, and flowers have been introduced recently to the community shared agriculture model. Still, vegetable production is the most common type of CSA (Schnell 2007). The consumers may also have a choice in the size of share they purchase. Usually a large share feeds a family of four, while a small share is enough to feed an adult couple (Ferris and Behmann

1994). Winter community shared agriculture, consisting mostly of root crops, is becoming more prominent in southern Ontario.

Community shared agriculture presumes that both the farmer and consumers share the benefits (crop output) and risks (various unpredictable aspects of growing food) of the season. In other words, members pay the real cost of food production (Ferris and Behmann 1994; Fieldhouse 1996; Cone and Myhre 2000; Krug 2004; Schnell 2007). If there is crop failure, the consumers lose on their initial investment. If there is an abundance of crops, the consumers gain. The cost of a share takes into account the costs of inputs and labour and the living wage of a farmer, although in the original model of community shared agriculture, the farmer and shareholders shared all of the risks. Allen (1999) believes the risks are transferred mainly to the consumers as they are not guaranteed to receive food on a weekly basis, but the farmer is still guaranteed a living wage. Schnell (2007), on the other hand, argues that community shared agriculture farms now operate as "subscription farms," where the fee that the shareholder pays up front guarantees a certain amount of food per week, rather than a share of the total harvest. Here, it is the farmer who takes all the risk as shareholders expect a certain amount of food per week. Schnell (2007) observed that many farmers purchase produce from other farmers to guarantee their shareholders will receive the expected amount of produce. The unequal distribution of risks between community shared agriculture farmers and shareholders raises issues of social justice in the CSA model and counters the initial reasons for developing CSAs. The idealistic model of sharing risks and benefits is not as prominent as it was in the original model of community shared agriculture.

Not all operations are certified organic, but most community shared agricultural operations are very aware of the effects of farming on the environment and therefore usually practise ecologically sustainable methods of farming. Hall and Mogyorody (2002) suggest that the large number of CSAs adds to the strong organic community in Ontario. Zsolnai argues that community shared agriculture is based on the notion of finite natural resources and operates the scale and profitability of the operation in relation to that notion (2002, cited in Feagan and Henderson 2009). Membership in CSAs challenges unlimited growth, unrestricted consumption, and merciless economic rationality (DeLind and Ferguson 1999). Genetic diversity is also promoted by growing a wide variety of crops. Some community shared agricultural producers save their own seeds or join groups or organizations that focus on preserving diversity of ecological or organic seeds (Fieldhouse 1996). Community shared agricultural operations also create a sense of community and educate the shareholders and the public about various forms of farming and their positive and negative contributions to the environment, health, the local economy, and social interactions.

Some aspects of the CSA model, including the cost of participation and location of distribution, may exclude some community members to some extent such as immigrants and poor people. Community shared agricultural operations are not as prevalent in the United States within poorer areas, and whites, Hispanics, and Asians are more likely than African Americans—who are among the poorest Americans on average—to be members of CSA operations (Schnell 2007).

Fieldhouse (1996) claims that community shared agriculture farmers view themselves as stewards of the land rather than just food producers or landowners. Cone and Myhre (2000) suggest that community shared agricultural operations are also capable of solving some problems posed by modernity. These problems include fragmented production, processing, and distribution of food that have increasingly disconnected consumers from producers. Community shared agriculture helps to reconnect producers with consumers, increasing consumers' awareness of where their food comes from and how it is produced. Although most CSA operations share a common structure, the locality and members should be taken into consideration during the development of the structure of a CSA (Bruno 1994; Hall and Mogyorody 2002; Schnell 2007). In fact, most community shared agricultural operations arise out of concerns about industrial agriculture and the global food regime, so CSAs are often associated with progressive, leftist political orientations (Schnell 2007).

North American Statistics about Community Shared Agriculture

While the first American community shared agricultural operation began in 1985 in western Massachusetts (Cone and Myhre 2000), by 2007, there were over 12,549 American CSAs (Community Shared Agriculture 2009), though in 2005 Schnell (2007) claimed there were only 1,700. Cuddleford (2003) suggests that direct sales, including community shared agriculture, of organic food may be the answer to saving the family farm in North America. Direct sales of local organic food may provide the farmer with up to 80 cents of each food dollar, whereas a conventional farmer participating in the wholesale market receives about 19 cents of each food dollar (Cuddleford 2003).

There were about five hundred community shared agricultural operations in Canada in 2002 (MacAdam 2002, cited in Haranandani 2009). To date, the Ontario CSA Farm Directory has approximately 250 registered CSA farms in Ontario alone. Community shared agricultural farms vary in size in terms of acres and number of shareholders. Ferris and Behmann (1994) mention the Armtree CSA farm near Mitchell, Ontario, which grew food for three families on a hectare of land, and the Weins' Twin Creeks Shared Farm at St. Adolphe, which grew food for 180 families on 25 hectares of land. However, Hall and Mogyorody's

(2002) study illustrates that while larger farms in Ontario are now more likely to have a community shared agricultural operation, most CSAs are based on smaller-scale farms.

Most CSAs in Canada are also located near a large city (Schnell 2007; Haranandani 2009) as operating one far from an urban centre and potential shareholders may be very difficult. On average, a community shared agricultural farm in southern Ontario grows over twenty varieties of vegetables (Ferrris and Behmann 1994). Krug (2004) claims that most shareholders prefer heritage crops as well as diversity in the foodstuffs they receive. In addition to satisfying the consumers with a diversity of food, the wide range of crops also contributes to the sustainability of the shares as it reduces the failure of all crops, thereby minimizing risk and uncertainty for shareholders.

Community Shared Agriculture as an Alternative to the Corporate Food Regime

Since community shared agricultural operations reached Europe and North America, they have been considered an alternative to the industrial food system (Fieldhouse 1996; Schnell 2007; Feagan and Henderson 2009) and as a tool to fight against the dominant global food system (Fairholm and Geggie 1998; DeLind and Ferguson 1999; Cone and Myhre 2000; Brehm and Eisenhauer 2008). DeLind and Ferguson (1999) believe members choose to participate in community shared agriculture to resist the dominant food system. Some community shared agriculture advocates believe that the large amount of food imports disrupts the development of food sovereignty in Canada. Community shared agriculture, on the other hand, contributes to the development of food security and sovereignty. Community shared agricultural operations offer a different form of economic participation than the dominant, industrialized capitalist system, but CSAs pale in comparative size to mainstream agriculture. They attempt to re-localize the economy by promoting the production of local goods that are valued by the local community (Fieldhouse 1996; Fairholm and Geggie 1998). The money paid to the producer stays in the local economy, strengthening many elements in that community, not just the CSA. Impersonal economic exchange is made more personal in community shared agricultural operations due to the direct interaction between the food producers and the food consumers. The direct interaction also minimizes any disruption that can take place beyond the control of the members and local community. On the other hand, if something goes wrong, such as a crop failure, there is no alternative system upon which the local community can depend other than the dominant industrial food system (Fieldhouse 1996).

Locavores also tend to like community shared agriculture whereby farmers are supported by a group of consumers who buy their produce at a central location, at the farms, or even have the produce delivered to the consumers. Occasionally consumers find that their produce has some insect damage, or they must pick it up from the farm at an inconvenient time, or they experience a glut of commodities they can't consume. However, most of the time, community shared agriculturalists have learned to manage these challenges quite well, such as, for example, using seed management systems and even freezing perishable crops for use after the growing season. Though CSAs do shorten the value chain by removing the intermediaries between the farmers and the consumers, thereby allowing consumers to get to know the farmers producing their food, sometimes consumers have learned that they do want intermediaries between themselves and the producers, assuming it makes accessing their food more convenient. If so, they may sacrifice freshness and community embeddedness. Community shared agriculture, like farmers' markets, enhance the community's embeddedness by creating "the context for closer social ties between farmers and consumers, but remain fundamentally rooted in commodity relations" (Hinrichs 2000, 295).

Whereas the dominant industrial food system is widely viewed as separating people from one another, from their community, from the seasons, and disrupting communities by separating food consumers from producers, community shared agriculture has created opportunities for participation with other members, their community, and the land. The values of these relations are placed against simple economic values. Fieldhouse (1996) claims that community shared agricultural operations create a different type of family farm, where instead of blood ties, the family is connected by common interests and passion for healthy food and a healthy food system. Because many community shared agriculturalists rely on urban centres for shareholders, CSAs also act as mediators between the rural and urban populations. Ostrom (1997, cited in DeLind and Ferguson 1999), Cone and Myhre (2000), as well as Hall and Mogyorody (2002) suggest that community shared agricultural operations can be considered a type of social movement. This cohesion can also provide opportunities for producers to work together to develop a strong community shared agricultural system, one that can deal with crop failures and other unpredictable risks (Fieldhouse 1996), and make culturally appropriate food available to the diverse ethnicities in Canada.

Changing Demographics and Ethnocultural Vegetables in Canada

Before exploring the potential that greater use of community shared agricultural operations holds for expanding the production and consumption of ethnocultural vegetables, it is important to contextualize CSAs with respect to Canada's changing demographics and ECV demand. The Canadian population is becoming increasingly diverse. In 2011 there were 1,567,400 South Asians,

1,324,700 Chinese, and 945,700 Canadians identifying themselves as black (Statistics Canada 2013c). Statistics Canada's *The Daily* (2013a) said that "Most of the 1.2 million immigrants who arrived in Canada between 2006 and 2011 settled in a Census Metropolitan Area (CMA)." Sixty-two percent of them settled in Toronto, Montreal, and Vancouver. These groups require food that they are familiar with based on religion, culture, or habit.

In addition to home gardening and community shared agriculture, there are novel models of distribution that enable and improve immigrant populations' access to fresh ethnocultural vegetables at more reasonable prices. Models such as Foodshare's The Good Food Box and The Mobile Good Food Market recognize that some neighbourhoods and members of the community face economic and geographic barriers, amounting at times to food deserts, limiting their access to fresh and ethnoculturally appropriate food. With the help of the non-profit sector, The Good Food Box and The Mobile Good Food Market can provide its members with high-quality, fresh, local fruits and vegetables at prices lower than in retail markets.

About 22.3 percent of participants interviewed in Adekunle, Filson, and Sethuratnam's (2010) study grew fuzzy melon, okra, bitter melon, basil, Chinese mint, smooth amaranth, African eggplant, and other vegetables in their backyards. Vegetables such as callaloo (smooth amaranth), bok choy, Jamaican pumpkin, and various squash and gourds have been grown in the Waterloo region (Gunst et al. 2010). Lack of access to backyard gardens and community gardens in cities limits the amount of ethnocultural vegetables residents can grow (Gunst et al. 2010). However, backyard gardening does not fulfill the demand for ethnocultural vegetables discussed in Chapter 2. This leaves a lot of potential for local food producers to meet the demand. Although growing conditions for these vegetables are not ideal in Canada, a large number of them, especially Chinese vegetables, have been grown in Holland Marsh, Ontario, for more than forty years (Davidson 2012). Adekunle, Filson, and Sethuratnam (2009, 2010, 2012) believe that despite the short growing season and climate in Canada, Canadian producers and the Canadian economy can largely benefit from growing ethnocultural vegetables locally.

Ethnocultural Vegetables and Community Shared Agriculture

As a result of the climatic conditions in Canada, some believe certain ethnocultural vegetables such as bitter melon and okra will not grow well in Canada and that ethnocultural vegetables in general will not grow organically (Davidson 2011). However, there are a large number of organic as well as small-scale producers who have identified the demand for ethnocultural vegetables many years ago and have successfully cultivated okra, bitter melon, a variety of eggplants, callaloo, bok choy, baby bok choy, Chinese greens, and other ethnocultural vegetables for many years. The model of community shared agricultural operations allows farmers to test the adaptability of various vegetables without risking their business. The model also allows them to test various marketing methods. Many claim that marketing ethnocultural vegetables is very different from marketing traditional produce (Adekunle, Filson, and Sethuratnam 2010, 2012; Davidson 2011; Daynard 2016). Specific vegetables have to be marketed in different ways to different cultural groups. The community shared agriculture model also allows producers to receive immediate feedback, which allows producers to gain a better understanding of which vegetables are popular with different groups and how to market them. Producers can also learn about the prices consumers are willing to pay for these vegetables.

Because many people garden in their backyards and in community gardens to help meet their demand for ethnocultural vegetables, many community members already participate in the alternative food system and value local, fresh, quality produce. Participation in the alternative food system and value for fresh produce suggests that many of these gardeners will be willing to purchase via the community shared agriculture model. The CSA model is flexible enough to adjust its marketing strategies to specific ethnic groups and to cultivate a variety of ethnocultural vegetables. The direct interaction with consumers and the mutual learning that takes place in the CSA model will likely encourage the identification of substitute crops for those vegetables that cannot be grown profitably in Canada. For example, elongated Asian eggplant is a great substitute for Italian eggplant. This type of knowledge is acquired in an arena that allows for interaction between various cultural groups and the exchange of various produce. The CSA model allows for this type of interaction and exchange to take place holistically, proficiently, and resourcefully. Thus the community shared agricultural model has the potential for creating a learning space that leads to an enabling environment for the development of appropriate production and sustained availability of fresh ethnocultural vegetables.

Policy Implications

In the ideal community shared agricultural model, members create strong relationships with their farmers and other local community members based on trust regarding the quality of the food and the farming practices on local farms. The members also have a long-term interest in working with the farmers and seeing the farms succeed economically. The farm is not only a source of fresh food, but also a tool for community building.

The dimensions of the relationship between producers and ethnic populations specifically have not been examined. Whether community building is a motivating factor for ethnic members to join community shared agriculture is unclear. However, there is a noticeable increase of ethnic CSA producers in southern Ontario and newcomers to Canada, and relatively more recent ethnocultural Canadians may join CSAs not only to access fresh, local, and familiar food, but also to connect with their community members and the wider community. These relationships expand consumer awareness about value of CSAs that goes beyond the dominant economic system.

Feagan and Henderson (2009) argue that this is somewhat unusual economic behaviour. Food is associated with more than monetary value and is re-embedded in a broader social, environmental, and cultural fabric. Growing ethnocultural vegetables in a community shared agriculture model can hold even greater cultural value for ethnocultural consumers as it helps them retain their traditions and culture and pass them on to future generations. Brehm and Eisenhauer (2008) also suggest that this strong relationship between the farmer and community facilitates a willingness to pay higher prices for food. This, along with the notion that ethnic populations equate freshness with quality and nutritional value, supports the argument that ethnic populations are willing to pay more for locally grown ethnocultural vegetables.

Many studies have focused on examining the motivations behind joining a community shared agricultural operation (DeLind and Ferguson 1999; Cone and Myhre 2000). Cone and Myhre's (2000) study suggests concern for the environment and desire for fresh and organic food are the dominant reasons for joining a CSA. Other studies argue more specifically that environmental concern is the main reason for joining (Kane and Lohr 1997; Ostrom 1997, cited in Brehm and Eisenhauer 2008). For these researchers, support for local food, knowing how and where the food is grown, and the desire to eat produce in season were secondary reasons. Their study suggests that doing something with a community, an opportunity to attend festivals, and price received low ratings (Fieldhouse 1996; Cone and Myhre 2000). Kane and Lohr (1997) and Ostrom (1997, cited in Brehm and Eisenhauer 2008), on the other hand, argue that community involvement is one of the greatest motivators for joining a community shared agriculture. However, as income rises, the other motivating factors become less significant (Brehm and Eisenhauer 2008).

Cone and Myhre's (2000) study also shows that many members were dissatisfied with receiving weekly food that they did not choose, especially when the produce was unfamiliar to them. Many members were not familiar with kohlrabi, kale, or rutabaga. Cooking, preparing, storing, and eating unfamiliar food were very challenging for the members and change their routines substantially (Cone and Myhre 2000). Kerton and Sinclair's (2010) research, on the other hand, suggests that receiving and learning how to cook unfamiliar food helped shareholders develop their knowledge about their local food system, including which vegetables and fruits can be grown in their areas as well as the concept of seasonality. As Cone and Myhre (2000) illustrate, the variety and quantity issues played a substantial role for why members did not renew their community shared agriculture memberships. Although unfamiliar food may act as obstacles to participating in CSAs for some, it is probably a motivator for others. If CSA producers market their ethnocultural vegetable produce accordingly, they may attract a larger number of ethnic communities. The number of community shared agricultural members from ethnic communities and their motivations are unknown and thus it is not clear how they would react to the variety of vegetables available in the CSA model. There is also a need for public awareness on the meaning and benefits of participating in community shared agriculture. At present, it seems a lot of stakeholders, and especially consumers, are not aware of how the CSAs operate.

Community shared agricultural operations have been criticized for not fully sharing the risks between farmers and consumers, especially regarding insect-damaged produce, the occasional lack of convenient drop-off locations, and produce gluts or shortages (Desrochers and Shimuzu 2012). These locavore critics and others argue that by eliminating the intermediaries between farmers and consumers, CSAs miss out on the value that wholesalers and retailers add within the supply chain. This view notwithstanding, some consumers and farmers are happy to do without these intermediaries while also gaining face-to-face contact.

It appears some members want to support the philosophy of the CSA, but they do not see it as necessarily very practical nor does it suit their lifestyles. It is difficult for consumers to transition to farm members. It is the farmers' role, Cone and Myhre (2000) suggest, to help members make that transition by creating a farm and rural experience for the members with whom they associate and identify. However, Hall and Mogyorody's (2002) study illustrates that community shared agriculturalists themselves believe their model is difficult and impractical. They believe in the benefits of CSAs, but do not think it is a long-term solution to the problems that CSAs attempt to address. Ostrom (1997, cited in Brehm and Eisenhauer 2008) claims that membership in community shared agriculture not only increases consumer awareness about food quality, health, and community sustainability, but has also altered their overall consumption behaviour. By contrast, Cone and Myhre (2000) say that CSAs often do not provide an incentive for changing consumption habits.

Our view is that CSAs provide a very healthy opportunity for people to participate in the local food movement. Therefore, the government ought to make it as feasible as possible for consumers to choose this option. Ontario's new policy of promoting local food production and consumption should therefore support and encourage inclusive community shared agriculture and farmers' markets.

The Way Forward

Bloom and Hinrichs (2010) think that as demand for local food increases, local food marketing and distribution will face coordination problems. Some authors suggest that the local food system may have to piggyback on the dominant and conventional food system rather than move toward direct marketing. As mentioned earlier, there are alternative models of distribution for local and fresh vegetables. In models such as The Good Food Box and The Mobile Good Food Market, food passes through secondary hands, but the models maintain the values of direct marketing. They support local farmers and help distribute fresh food to as many people as possible, but some of the benefits of direct marketing, such as the strong relationship between farmers and shareholders in community shared agriculture, are not included.

Bloom and Hinrichs (2010) propose an expanded value chain as a model for restructuring local food systems. This would allow the local food system to operate on a larger level, but maintain some of the social and environmental aspects of direct marketing. Increasing access to technology has allowed for the development of unique food distribution systems. There are numerous online food distribution systems around Canada. Consumers order food online, which is delivered to them, usually on a weekly basis. Again, the food passes from local farmers to the distributors to the consumers. Although this is not direct marketing, many of these companies directly support local farmers, are engaged in various community and charity work, and educate their consumers about local seasons, local produce, and the farmers who grow the food. These companies are for profit but create positive change in their communities, environments, and local food systems. Many of these companies also offer ECVs such as bok choy, callalloo, and varieties of eggplant. These online markets can potentially be more accessible than farmers' markets and community shared agriculture, though they don't usually have the benefit of social interaction.

However, aligning local food and direct marketing with the conventional food system may increase conflict within the local food system as members of the community may begin to question the goals and meanings of local food. This may create more concern as the role of online distribution companies in supporting local food increases. The internal conflict may result in political weakness, as it did in California with organic food (Hall and Mogyorody 2002). On the other hand, the conflicts may motivate the local food system to build stronger relationships and interaction as well as strengthen their system via re-evaluation.

Direct marketing is essential to the alternative food system, which includes community shared agriculture. Hall and Mogyorody (2002) think that the marketing system of local producers is their strength and can be utilized as a political tool. Mount (2012) believes that the value of the local food system depends on the elimination of middle men. Mount (2012) discusses the short food supply chains in Europe and argues that value is added as a result of that short chain, rather than through processing produce. Dismantling the direct relationship between farmer and consumer can be destructive to the local food system (Hall and Mogyorody 2002; Mount 2012). Accountability, trust, and confidence that stem from direct marketing, such as community shared agricultural operations, may be lost if the local food system were to scale up (Mount 2012). Producers can use their direct relationship with consumers to contest any effort by the government or agribusiness sector that aims to alter local or organic farming. Although neither the government nor agribusiness has given support for local ethnocultural food, the produce is increasingly visible in supermarkets in southwestern Ontario and elsewhere. This suggests that the popularity of local food has already been largely established.

Nonetheless, some question the market strength of the local food consumer and the alternative marketing system. Some community shared agriculturalists have long waiting lists and the increasing availability of local and ethnocultural food in both large chain and small stores may discourage people from direct marketing models. Taking into consideration the more than $61 million demand for ethnocultural vegetables per month in the GTA alone and the growing conditions in Canada, it is unlikely that the community shared agriculture model alone will meet that demand. Models such as The Good Food Box and online distributors can fill that gap where the CSAs cannot. Moreover, the CSA model can help overcome the scarcity of fresh locally produced ethnocultural food in a space where social interaction is encouraged.

Some authors point to the notion that the local food system is flexible and is capable of creating hybrids (Hall and Mogyorody 2002; Mount 2012). The wide range of motivating factors for participating in community shared agriculture may suggest that there is flexibility in what consumers are expecting from an alternative food system. Some motivating factors—such as the value for local, high-quality produce that is part of a larger environmentally friendly system of food production—are unlikely to change resulting in shareholders' purchasing produce from mainstream stores. In Hall and Mogyorody's (2002) view, the community shared agriculture model appears to be flexible enough to adapt to the changing demands of consumers, their communities, and wider society.

Finally, some authors suggest that community shared agriculture is not as accessible by underprivileged populations as they should be (Hinrichs and Kremer 2002; Kneen 1991). This is one of the issues FoodShare attempts to address via The Good Food Box and The Mobile Good Food Market. Brehm and Eisenhauer (2008) argue that further examination on the composition of members, their motivations, and how CSAs may increase social capital could be undertaken to improve the participation rate of various segments of the population. Canada cannot count on a strong food culture to steer the food industry to a local food supply, compared with Europe, which has more deeply rooted food cultures. In Canada, it is up to the public to create a demand for a local food system (Friedmann 2007) and for the government to create an enabling environment that will inspire and motivate Canadians to participate in alternative ways of accessing food, such as community shared agriculture. As Schnell (2007, 557) argues, "This move to the local is a logical counter-movement for those with anticorporate and antiglobalization ideals."

CHAPTER 7

Growing and Consuming Our Way to a Healthier People and Economy

Glen C. Filson and Bamidele Adekunle

Contradictions Affecting Ethnocultural Vegetable Production and Consumption

The issues affecting people's access to ethnocultural vegetables are political and economic insofar as they challenge production and marketing, which has such a deleterious effect on the environment, super-exploits vegetable workers, and leaves consumers wanting their culturally preferred, fresh vegetables. In this book, we distinguished the corporate food regime from the local food and community food sovereignty movement. The latter advocates much shorter value chains, incurring fewer food miles, and providing more producer-consumer contact, thereby partially mitigating the metabolic rift between the global capitalist commodity chains and the environment (Foster et al., 2010). As Foster et al. (2010) put it:

> An irreparable rift (rupture) emerged in the metabolic interaction between humans and the earth, one that was intensified by large-scale agriculture, long distance trade, massive urban growth, and large and growing synthetic inputs (chemical fertilizers) in the soil. (Foster et al. 2010, 124)

Shortening the ethnocultural food value chain, improving food sovereignty and security, producing more organically and reducing exploitation wherever possible in the production and consumption of food will help mitigate the metabolic rift between society and nature.

We identified the contradiction between the cheap industrial diet with its convenient, processed food despite people's need for healthy but more expensive culturally preferred fruits and vegetables. These vegetables protect people from many diseases of the heart, cancers, diabetes, and many other ailments, thereby reducing strain on our health care system. Ethnocultural vegetables are also fresher and more nutritious when grown locally.

Elaborate social class contradictions within the corporate food regime exist between the chain store purchasers of tropical and semi-tropical fruits and vegetables from peasants, small-operation farmers, (who have access to some land but must support their farming through their off-farm jobs), or selling their labour for a wage in fields owned by those buyers or dependent owners. In Canada there are other contradictions between the major supermarkets (represented by the Retail Council of Canada), which control more than 70 percent of sales and the smaller businesses who are represented by the Canadian Federation of Independent Grocers.

Existing contradictions between industrial agriculture and both small-operation producers and exploited workers, between financialized global supermarkets and small ethnic stores, between globally sourced, stale vegetables and consumers wanting healthy, fresh produce all contribute to the unsustainability of the present social class contradictions within ethnocultural supply chains. A huge contradiction also exists between the corporate food regime's buyer-driven, vertically controlled system of ethnocultural vegetable provision and the various local food provisioning alternatives, including foreign/local large producers versus small-operation local growers of ethno-cultural vegetables.

A new balance between the long- and short-distance ethno-cultural vegetable chains is needed to overcome the contradiction between the corporate food regime's predominance versus the alternative, local, and mainly organic food movement. This new approach must have the potential to make an abundant supply of fresh ethnocultural vegetables available at better prices while vastly improving the working conditions, job safety, and access to health care available to the temporary foreign workers who do much of the work producing these vegetables. While capital moves easily from place to place, labour is far less mobile, which helps to reinforce capitalist access to cheap, easily exploitable labour power throughout the global vegetable value chains.

At present not enough is known about ethnocultural vege-table cultivars; their growing season requirements in Ontario or Canada as a whole; access to seeds, transplantation, nutrition, fertilizer requirements, and suitable spacing for these crops.[1] Minor-use pesticides are almost completely unavailable for the production of these vegetables, which is necessarily very labour intensive. Clearly climate change continues apace with some scientists predicating a rise in temperature ranging from a min-imum of 1.5°C in the most optimistic forecasts to as much as 4°C 2100 (IPCC 2016). If so, as time goes by, despite our best efforts to mitigate climate change, we will be able to grow more and more ethnocultural vegetables in Canada.

Not only is there already a significant Chinese horticultural class growing these vegetables in Holland Marsh, north of Lake Erie, and in the Niagara region, increasingly there is a readily avail-able multi-ethnic workforce, some of whom are becoming petty commodity producers on rented or even purchased southern Ontario land. As the demand for their culturally appropriate vege-tables continues to grow, supermarkets are expanding their supply of these vegetables, often presenting them in separate ethno-cultural fruit and vegetable stalls. Supermarkets are working more with local producers willing to produce, harvest, and package the vegetables to supermarket specifications.

While some farmers of European descent are experimenting with ethnocultural vegetable production, more are still waiting to learn about which vegetables grow best in which soils and at what cost of production. Commercial horticulturalists are also waiting for more registered pesticides to be approved by Health Canada before taking the risk of planting in the absence of assured crop protection (Health Canada 2013).

Ways of Propagating People's Culturally Preferred Vegetables

There is still a strong need to bridge the cultural gap between European-descent horticultural producers and relatively recent multi-ethnic immigrants, especially if we are to convince Canadian commercial farmers to grow more ethnocultural vegetables. People's right to access their culturally preferred food should be enshrined in our provincial human rights codes. When people have food sovereignty, they are much happier with their integra-tion. This should be seen as an essential path to incorporating immigrants and refugees. Canada's openness to receiving 31,000 refugees from wartorn Syria in 2016 was reminiscent of our open-ness to the integration of the Vietnamese boat people in the late 1970s, many of whom are now prominent in business, profes-sional, and other important walks of life. There has been greater emigration from many countries and immigration to other coun-tries. If they can access their basic human right to food sover-eignty, all Canadians benefit and it increases everyone's access to a diversity of cuisines.

These vegetables are being made more accessible in farmers' markets, community shared agriculture, and food hubs like western Toronto's The Stop, which organizes, distributes, and markets local food by connecting consumers with local producers. Their priority remains for the food to be as fresh and nutritious as possible. But much more agricultural extension needs to be done in order to alert the largely British and Dutch Canadian producers to the demand for these ethnocultural vegetables, including ones that can be grown most readily in Canada.

The expectations for Ontario's extension system, or rather, the systemic intercultural form of Knowledge Translation and Transfer (KTT) as OMAFRA now characterizes agricultural extension, is to promote the conditions for profitable local production of ethnocultural vegetables. So far OMAFRA has been modestly successful in (a) promoting the local food movement and (b) explaining the large ethnocultural vegetable demand to seed companies, farmers, assemblers, wholesalers, retailers, and consumers. Much more can be done to promote the crossover potential and obvious health benefits that these ethnocultural vegetables provide to Canadians as a whole thereby strengthening multicultural, intercultural Canada.

In this regard, ECVOntario has facilitated an exchange of ideas about how best to do this among those who are part of the ethnic vegetable value chain.[2] People are most likely to gain awareness of these products through print media, but we are also using social media to promote recipes and the benefits of eating these vegetables. Toronto Public Health and the Toronto Food Policy Council can use their websites to post information about how these vegetables can improve people's health.

Right now the most substantial local production and marketing of these vegetables is being done by Chinese Canadians, many of whom are relatively isolated from or ignored by those dominating the local food movement (Gibb 2011). According to Gibb and Wittman in the lower British Columbia mainland,

> An older network [of the local food movement] consists of roadside stores and greengrocers supplied by Chinese-Canadian farmers. A newer, rapidly expanding network includes farmers' markets and other institutions publicly supported by the local food movement. Both networks are "local" in that they link producers, consumers, and places; however, these networks have few points of intentional connection and collaboration. (2013, 1)

This probably applies to southwestern Ontario as well. Much more, therefore, needs to be done within the overall movement to integrate the parallel Chinese Canadian local food leaders as well as South Asian and Afro-Caribbean local food leaders with the mainstream, mostly European Canadian movement.

Farmers' markets and community shared agriculture have functioned almost entirely on a voluntary basis and have never had the kind of government support in Canada that mainstream export-oriented agriculture has such as the federal and provincial contributions to farmers' agricultural safety net. Earlier we proposed tax credits for ethnocultural vegetable producers, but were criticized by a few OMAFRA members, who reminded us that the government should not directly subsidize producers as this would put us at odds with World Trade Organization rules. Some economic incentives may still be needed, however, such as ramping up locally produced ethnocultural vegetables (and probably regulated marijuana) to replace existing unhealthy tobacco production.

Our socio-cultural work in ECVOntario complements the socio-economic and horticultural research of the Ontario Fruit and Vegetable Growers' Association, Vineland Research and Information Centre, the Simcoe Research Station, and the

University of Guelph's Muck Research Station in the Holland Marsh on ethnocultural vegetables. The Ontario Ministry of Agriculture, Food and Rural Affairs has done a lot to encourage farmers to try growing ethnocultural vegetables. We have found, however, that the cultural barriers between the farmers of European descent and the multi-ethnic consumers of non-European descent cannot be ignored. This cultural gap partly explains why farmers have been relatively unaware of the growing demand for these vegetables from the GTA and the northeastern United States. This also partly explains why producers have not taken more advantage of this growing market segment. Also, not enough is yet known about how distributors and processors within the ethnocultural vegetable value chains can benefit from their local production and sale.

We have sought to expand producers' knowledge networks, repertoire, and improve agricultural and farming practices especially geared to local food production and processing, but because most horticultural organizations like the Ontario Fruit and Vegetable Growers' Association embody cultural systems and are engaged with other systems, contradictions among these systems remain to be precisely identified and transformed.

Canadian immigrants are willing to pay a premium for fresh, locally produced vegetables (Adekunle, Filson, and Sethuratnam 2010). These vegetables will have their full health and nutraceutical benefits in comparison to imported/frozen produce. Carter-Whitney and Miller (2010) of the Canadian Institute for Environmental Law and Policy argue that the legal, regulatory, and international agreement barriers to local food production and marketing are surmountable. Over the past few decades, food procurement and distribution systems have consolidated into a few large corporations, severely restricting the purchase of local food. Canada has found itself bound by two free trade agreements, NAFTA and the WTO, which are designed to remove most obstacles for trade between countries. As we write, U.S. President Trump wants to renegotiate NAFTA to benefit Americans and he has walked away from ratifying the Trans-Pacific Partnership (TPP). It is unclear at this time whether the European Union will ratify the Canada-Europe Comprehensive Free Trade Agreement (CETA), but it is clear that these agreements generally privilege multinational corporations' rights over local rights regarding matters of investment decisions. For instance, assuming we proceed with CETA, "The chapter on Government Procurement widens corporate penetration into governments at every level by generalizing 'national treatment' and prohibiting 'offsets' defined as 'any condition or undertaking that encourages local development'" (Rossman 2016).

Instead of adopting these new agreements and continuing with the old free trade rules, Canada needs to find a way to review and make adjustments to these agreements in the best interests of its people (Carter-Whitney 2008). She recommends that the Ontario government work with the Ontario Food Terminal Board to promote local food and use other market-based initiatives that would enhance the purchase of locally grown food. She also suggests that food labelling be improved to make the country of origin clearer in addition to using local food labelling to help consumers identify fresh, locally produced food. As good as labelling is for traceability, it is expensive for small farmers and it can be an economic disincentive unless costs can be shared with the government.[3]

A number of challenges and risks must be addressed in order for the expansion of the ethnic food processing to succeed. The ethnocultural vegetable producer in Ontario must use effective marketing strategies to connect with not only the ethnic population in the Greater Toronto Area but similar ethnic consumers in the northeastern United States. It is also clear that consumers value authenticity of taste, though second-generation immigrants are likely to be more accepting of Western modifications as the result of their greater acculturation. It can be difficult to achieve authentic taste in Canada, as authenticity implies obtaining the food from the country of origin, growing it locally, and using traditional preparation methods (WCM Consulting 2008).

Stahlbrand (2006) observes that consumers need to be informed about the farmers in their area who produce fresh food for retail. She notes that there are a number of obstacles to sustainable mainstream food production, including the extent to which they use fossil fuels, create greenhouse gases, degrade the soil and pollute the water. She feels that a local food system would improve sustainability by using financially profitable methods for the farmers and local communities while at the same time protecting human health and the environment.

Lessons Learned

Cultural groups tend to have a lot in common regarding food. For instance, okra has significant cross-cultural impact and is in high demand among the major recent immigrants to the GTA. FarmStart has made progress enabling some of the most recent immigrants to grow ethnocultural vegetables on incubator farms, but we need policies that will alleviate the barriers to local ethnocultural vegetable production such as by innovative leasing arrangements so that the marketing of locally produced produce is enhanced through supermarkets, ethnic stores, farmers' markets, and community shared agriculture. In the past, there was little support from governments or agribusinesses for locally produced ethnocultural food, but the Ontario government's passing of the Local Food Act (2013) is a positive step forward.

Canada's short cultivation period, pests and diseases, lack of seeds and pesticides, higher labour costs, and relative lack of government support, at least until recently, for farmers attempting to cultivate ethnocultural vegetables locally, has held us back. However, the present lack of knowledge about production methods and poor understanding of how our farmers can provide more ethnocultural vegetables need to be addressed especially in the light of global warming and future prospects for these vegetables.

An even bigger challenge for small operators has been the high cost of land for newcomers attempting to produce their culturally preferred vegetables (as much as $20,000 per acre or more in Oxford County). This is obviously most significant for people with less capital, such as recent immigrants. Leasing land is one way around this problem, but more must be done to encourage non-whites with farming experience to produce for the ethnocultural vegetable market. At present, there are many municipal bylaws that make it difficult for farmers to lease small parcels of land because of concerns about land fragmentation.[4]

Progressive conservation authorities have made available land that they control for community gardens where ethnocultural vegetables are increasingly being produced. McVean Farm is one example in Brampton, thanks to the Toronto Conservation Authority. Other groups like Ignatius College in Guelph have also gone out of their way to provide community gardens and encourage community shared agriculture. Much more can be done by local governments, municipalities, conservation authorities, and generous private groups and citizens.

Well-developed, complementary ethnocultural vegetable value chains will be beneficial to people who are not necessarily from the same cultural groups. Such chains also encourage a better appreciation of other people's culture because though many perceive recent immigrants as different, they still have much in common with everyone else, especially in terms of food preferences. Acculturation usually starts with the acceptance of other ethnic groups' food, which may lead to better appreciation of other cultures' values and norms. Furthermore, positive acculturation will increase the demand for ethnocultural vege-tables. This has already happened to a substantial degree with Chinese vegetables, which is an indication of conscious and sub-conscious acculturation.

The retail sector of ethnocultural vegetables is hierarch-ical, relational, and complex. The pricing structure is ambigu-ous. We cannot categorically state that ethnocultural vegetables are cheaper in mainstream stores than ethnic stores because it depends on the produce, the cultural group that mostly consume it, the cross-ethnic impact, and whether or not it can be produced locally. Although some Afro-Caribbean stores are improving, at times they charge arbitrary prices because their market is not as developed as the Chinese or South Asian markets. Furthermore, grocery buyers are interested in different attributes when they shop for ethnocultural vegetables. The Chinese often choose to patronize stores where food is labelled in Mandarin or Cantonese, but often there are no English and/or French labels, which may also help sales if they are present.

Spaces where local ethnocultural vegetables can potentially thrive are presently too few. Partly this is because those most likely to seek healthy, fresh fruits and vegetables tend to be rela-tively educated, bourgeois, professional, and upper-working-class people who are disproportionately of European descent. Alternative agriculture such as farmers' markets, community shared agriculture, and gardens can increase the production of ethnocultural vegetables in Canada, but policy makers need to create an enabling environment and prescribe policies that will make such spaces more inclusive and affordable. Change is inevitable and cultural groups also assimilate and learn from other cultures once they continue to interact in the same space. In the future we suspect Canada will move toward a fusion of cuisines that are more culturally appropriate for all cultural groups, thereby promoting better public health.

Partly because many of the present contradictions affecting people's health and access to their culturally preferred vegetables is papered over by existing government policy, changes are needed to make the situation more transparent to overcome the most significant contradictions affecting the production, distribution, and consumption of ethnocultural vegetables. The contradiction between the primarily anglophone and Dutch Canadian farmers producing mainstream vegetables and the rising wave of South Asian, Chinese, and Afro-Caribbean immigrants seeking their culturally preferred vegetables is partly cultural, partly informational yet potentially a win-win situation for both. If seasonal workers from the Caribbean and Latin America can be treated better and eventually provided with the choice to become naturalized Canadian citizens, this will overcome the most important ethical dilemma and barrier to the production of ethnocultural vegetables.

More connections should be made between locally grown ethnocultural vegetables, not only with the Ontario Fruit and Vegetable Growers' Association, the Ontario Processing Growers' Association, Fresh Ontario Greenhouse Growers, the Ontario Greenhouse Marketers Association, the Ontario Horticultural Association, and the Landscape Ontario Horticultural Trades Association.

Federal, provincial, and municipal policies are needed to make it easier for locally produced ethnocultural vegetables to find Ontario markets through ethnic stores, supermarkets, farmers' markets, and community shared agriculture. Because production of ethnocultural vegetables does not guarantee market access, we also need to promote policies conducive to the local production, wholesaling, retailing, and alternative marketing of ethnocultural vegetables. Our studies of short-distance value chains for these ethnocultural vegetables have clarified the relative differences in seed distribution, costs of production, marketing, and consumption of imported versus local production. We hope that this will help shift the balance to greater local production. Identification of the links between different components of ethnocultural vegetable value chains has identified some policies that will enable farmers to benefit from this market sector, but much more is needed to enable people to access their culturally preferred food.

Development of minor-use pesticides for specific ethno-cultural vegetables should also be encouraged and developed as organic production is not for everyone, especially given the premium prices organic producers must charge. Education and communication of friendly policies are needed to overcome the cultural barrier between European commercial growers not interested in ethnocultural vegetable production and newcomers seeking their culturally appropriate food.

The prospects for alternative forms of agriculture such as local farmers' markets and community shared agriculture in the face of the present dominant corporate food regime are growing as awareness about the local food movement and its potential also grows, enhancing people's food sovereignty in Canada. Therefore, "food sovereignty is a counter movement expressing the central contradiction of the food regime and its circulation relations" (McMichael 2016, 651) globally. The market prices for these ethnocultural vegetables represents their exchange and use value, which change along with changes in their prices of production (the capital cost of growing them based on their labour, equipment, and transportation costs, plus an average rate of profit) between different seasons of the year.

Having determined that the production, distribution, and consumption of ethnocultural vegetables are spread across global, long-distance versus short-distance chains, we have advocated shortening these supply chains for consumers within the Greater Toronto Area and southern Ontario as much as is feasible. Members of the local food movement are partly motivated to provide an alternative to the corporate food regime thereby disrupting the disastrous metabolic rift between capitalism and the environment, mitigating climate change, and improving ethnocultural vegetable agro-ecosystem sustainability.

Notes

Introduction

1. The *Toronto Star* (October 4, 2016) reports that post-census statistics put the Greater Toronto Area's population at 6.63 million (including Burlington in the west and Clarington in the east) and the Toronto Region's population at 6.13 million.

Chapter 1

1. Modes of production entail specific production forms and types of surplus extraction such as slavery (masters' use of slaves), feudalism (landlords' collection of rents), and capitalism (capitalists' profit derived from unpaid surplus labour expropriated from workers). Each mode has particular forces of production (technology, labour quality), property connections, and social relations of production (that is, class structure). As Marx explained:

   > It is always the direct relationship of the owners of the conditions of production to the direct producers—a relation always naturally corresponding to a definite stage in the development of the methods of labour and thereby its social productivity—which reveals the innermost secret, the hidden basis of the entire social structure, and with it the political form of the relation of sovereignty and dependence, in short, the corresponding specific form of the state. (1959, 791)

Chapter 2

1. Our classification differs slightly from that of Aujla, who follows Henry in saying that "Generally, the category 'South Asian' refers to those who trace their ancestry to places including India, Pakistan, Sri Lanka, Bangladesh, Bhutan, Tanzania, Uganda, South Africa, and the Caribbean" (2000, 41). Those coming from Africa and the

Caribbean are treated separately in our study as Afro-Caribbean Canadians. Readers may wonder why Southeast Asian Canadians (Indonesians, Malaysians, Japanese, Vietnamese, Thais, Laotians, Burmese, Cambodians, etc.) are not discussed here. Certainly there are many Southeast Asians living in the GTA, but we only had the resources to interview members of the three largest ethnic groups as reported in Statistics Canada (2006b).

2. Japanese Canadians' removal to detention camps during the Second World War, despite their long residence, was another example of their horrendous treatment. Civil rights were not restored to Japanese Canadians until 1949 (Adachi 1976).

3. Upper Canada's Gov. John Simcoe's attempt to abolish slavery in 1792 was not entirely successful. In 1793 the Upper Canadian Legislature ruled that the slaves' children must be freed at the age of twenty-five, and no more slaves could be imported.

Chapter 5

1. According to Marzall, Filson, and Adekunle (2011, 26), "Mindscapes are defined by a combination of different elements, including perception, knowledge (encompassing conscientization and awareness), worldview, and values (priorities, expectations, preferences and motivations)."

Chapter 7

1. Vineland Research and Innovation Centre has discovered that "During Vineland's research trials, the highest yield was achieved when okra was planted on double rows spaced 180 cm centre-to-centre and 25 to 30 cm plant-to-plant in the row. Daily harvest of okra is necessary during peak production in the summer to ensure top-quality produce. During the fall, harvest slows to two to three times per week" (Daynard 2016, 4).

2. This occurred on February 17, 2012, at the workshop, Ethno-Cultural Vegetables Ontario: Challenges and Opportunities of the

Ethno-Cultural Vegetable Market. Well over one hundred representatives from Simcoe Research Station, Vineland Innovation and Research Centre, Ontario Food Terminal, Ontario Fruit and Vegetable Growers' Association, Ontario Greenhouse Growers Association, Greater Toronto Agricultural Action Committee, ethnic and mainstream store representatives, and farmers, including FarmStart's McVean Farm and many other farmers, came together to discuss ways of increasing the production and consumption of ethnocultural vegetables. At this event, members of the audience were shown the documentary *Old Food, New Land: Canada's Unexplored Niche Market, A Short Film*, produced by Adekunle, Filson, and Sethuratnam (https://www.youtube.com/watch?v=o6fzSYKzSUs; https://www.youtube.com/watch?v=y-EJ0NLqAmM). Social media platforms such as Facebook, Twitter, Blogger, and YouTube have also been and continue to be used by various members of ethnocultural vegetable value chains to discuss the issues regarding ethnocultural vegetables and the local food movement.

3. In January 2009 the Canadian Food Inspection Agency introduced new regulations that require products labelled "Product of Canada" to be 98-percent Canadian origin, while "Made in Canada" must have Canadian material and the labour used must be at least 51-percent Canadian (Stevens 2009).

4. Understandably, in recent decades municipalities have moved to block severances that enabled urbanites' homes to proliferate within farming communities, which has exacerbated tensions between farm and non-farm rural folks. But more could be done to allow severances of 5- or 10-acre plots, which could be sold to immigrant farmers who want to grow the ethnocultural vegetables they know how to grow and love to eat.

References

AAFC (Agriculture and Agri-Food Canada). 2007. *Special Features: Census of Agriculture Summary.* Ottawa. http://publications.gc.ca/collections/collection_2008/agr/A38-1-1-2007E.pdf.

Abdel-Ghany, M., and D. L. Sharpe. 1997. "Consumption Patterns among Ethnic Groups in Canada." *Journal of Consumer Studies and Home Economics* 21, no. 2: 215–23.

Adachi, K. 1976. *The Enemy That Never Was: A History of the Japanese Canadians.* Toronto: McClelland and Stewart.

Adekunle, B., G. C. Filson, and S. Sethuratnam. 2009. *Consumption of Ethnocultural Foods: A Research Guide.* Guelph: University of Guelph.

———. 2010. *Preferences for of Ethnocultural Foods in the Greater Toronto Area: A Market Research.* OMIF Technical Report. Guelph: University of Guelph.

———. 2012. "Culturally Appropriate Vegetables and Economic Development. A Contextual Analysis." *Appetite* 59, no. 1: 148–54.

———, executive producers. 2012. *Old Food, New Land: Canada's Unexplored Niche Market, A Short Film by ECVOntario.* Sponsored by OMAFRA, Agri-Food and Rural Link, FarmStart, Vineland Research and Innovation Centre, and the University of Guelph. https://www.youtube.com/watch?v=o6fzSYKzSUs; https://www.youtube.com/watch?v=y-EJ0NLqAmM.

———. 2013. "Immigration and Chinese Food Preferences in the Greater Toronto Area." *International Journal of Consumer Studies* 37, no. 6: 658–65.

Adekunle, B., G. C. Filson, S. Sethuratnam, and D. Cidro. 2011. "Acculturation and Consumption: Examining the Consumption Behavior of African Descendents in Canada." *Journal of Agriculture, Food Systems, and Community Development* 2, no. 1: 1–17.

Agriculture, Food and Rural Development. 2004. *Understanding Consumer Trends Can Present New Opportunities.* Agri-Processing.

Government of Alberta. http://www.agric.gov.ca/$department/
deptdocs.nsf/all/sis8735/$file/8735/pdf?OpenElement.

Ahumada, O., and J. R. Villalobos. 2009. "Application of Planning
Models in the Gari-food Supply Chain: A Review. *European Journal
of Operational Research* 195: 1–20.

Aitken, J. 2014. "Market Status of Okra, Chinese Long Eggplant and
Indian Round Eggplant in the Greater Toronto Area." MSc thesis,
University of Guelph.

Albrecht, C. 2014. "Reconnection in Local Food Issues: Purpose,
Practice and the Calibration of Value." MA geography thesis,
University of Guelph.

Alkon, A. H., and C. G. McCullen. 2011. "Whiteness and Farmers
Markets: Performances, Perpetuations. Contestations?" *Antipode* 43,
no. 4): 937–59.

Allen, P. 1999. "Reweaving the Food Security Safety Net: Mediating
Entitlement and Entrepreneurship." *Agriculture and Human Values*
15, no. 2: 117–29.

Anderson, A. B., and J. S. Frideres. 1981. *Ethnicity in Canada: Theoretical
Perspectives*. Toronto: Butterworths.

Apparicio, P., M. S. Cloutier, and R. Shearmur. 2007. "The Case of
Montreal's Missing Food Deserts: Evaluation of Accessibility to
Food Supermarkets." *International Journal of Health Geography* 6: 4.
https://www.ncbi.nlm.nih.gov/pubmed/17295912.

Asfaw, A. 2008. "Fruits and Vegetables Availability for Human
Consumption in Latin American and Caribbean Countries:
Patterns and Determinants." *Food Policy* 33: 444–54.

Asian Groceries in Toronto. 2012. www.blogto.com/grocery/c/toronto/
asian-groceries/.

Asp, E. H. (1999). Factors Affecting Food Decisions Made by Individual
Consumers. *Food Policy* 24, no. 2: 287–94.

Aujla, A. 2000. "Others in Their Own Land: Second Generation South
Asian Canadian Women, Racism, and the Persistence of Colonial
Discourse." *Canadian Women Studies* 20, no. 2: 41–47.

Avakian, A.V., ed. 1997. *Through the Kitchen Window: Women Explore
the Intimate Meanings of Food and Cooking*. Boston: Beacon Press.

Awolu, O. O., R. O. Osemeke, and B. O. Temilade Ifesan. 2016. "Antioxidant, Functional and Rheological Properties of Optimized Composite Flour Consisting of Wheat and Amaranth Seed, Brewers' Spent Grain and Apple Pomace." *Journal of Food Science and Technology* 53, no. 2: 1151–63.

Bäckström, A., A. M. Pirttilä-Backman, and H. Tuorila. 2003. "Dimensions of Novelty: A Social Representation Approach to New Foods." *Appetite* 40, no. 3: 299–307.

Baker, L. 2004. "Tending Cultural Landscapes and Food Citizenship in Toronto's Community Gardens." *Geographical Review* 94, no. 3: 305–25.

Balkissoon, D. 2016. "Migrant Farm Workers Deserve Better." *Globe and Mail*, September 19, A13.

Bannerji, H. 1993. "Popular Images of South Asian Women." *Parallelogramme* 11, no. 4: 17–20.

Basran, G. S. 1993. "Indo-Canadian Families: Historical Constraints and Contemporary Contradictions." *Journal of Comparative Family Studies* 24, no. 3: 339–52.

Bell, D. 2013. "The Business of Going Green: Agri-Environmental Certification in Ontario, A Multiple Account Benefit-Cost Analysis." MSc thesis, University of Guelph.

Bernstein, H. 2014. "Food Sovereignty via the 'Peasant Way': A Skeptical View." *Journal of Peasant Studies* 41, no. 6: 1031–63.

————. 2016. "Agrarian Political Economy and Modern World Capitalism: The Contributions of Food Regime Analysis." *Journal of Peasant Studies* 43, no. 3: 611–47.

Birch, L. L. 1999. "Development of Food Preferences." *Annual Review of Nutrition* 19: 41–62.

Bissoondath, N. 1994. *Selling Illusions: The Cult of Multiculturalism in Canada.* Toronto: Penguin Books.

Black, J., and J. Macinko. 2008. "Neighborhoods and Obesity." *Nutrition Reviews* 66, no. 1: 2–20.

Blackwell, R. D., P. W. Miniard, and J. F. Engel. 2001. *Consumer Behaviour,* 9th ed. Orlando, FL: Harcourt.

Bloom, J. D., and C. C. Hinrichs. 2010. "Moving Local Food Through

Conventional Food System Infrastructure: Value Chain Framework Comparisons and Insights." *Renewable Agriculture and Food Systems* 26, no. 1: 13–23.

Bodor, J. N., R. Donald, T. A. Farley, C. Swalm, and S. K. Scott. 2007. "Neighbourhood Fruit and Vegetable Availability and Consumption: The Role of Small Food Stores in an Urban Environment." *Public Health Nutrition* 11, no. 4: 413–20.

Bonacich, E. 1980. "Class Approaches to Ethnicity and Race." *Insurgent Sociologist* 10, no. 2: 9–13.

Bond, D., and R. Feagan. 2013. "Toronto's Farmers' Markets: Towards Cultural Sustainability?" *Journal of Agriculture, Food Systems, and Community Development.* http://dx.doi.org/10.5304/jafscd.2013.032.005.

Boothman, B. 2009. "A More Definite System: The Emergence of Retail Food Chains in Canada, 1919–1945." *Journal of Macromarketing.* Originally published online December 11, 2008. http://journals.sagepub.com/doi/abs/10.1177/0276146708327622.

Boyd, M., and M. Vickers. 2000. "100 Years of Immigration in Canada." *Canadian Social Trends.* Catalogue no. 11.008. Ottawa: Statistics Canada.

Brehm, J. M., and B. W. Eisenhauer. 2008. "Motivations for Participating in Community-Supported Agriculture and Their Relationship with Community Attachment and Social Capital." *Southern Rural Sociology* 23, no. 1: 94–115.

Briggs, J. 1985. "An Exploratory Study of Farmers' Choice of Crops in Central Sudan." *Transactions of the Institute of British Geographers* 10, no. 2: 170–80. http://www.jstor.org/stable/621821.

Bromley, T. V. 1974. "The Term 'Ethnos' and Its Definition." In *Races and Peoples: Contemporary Ethnic and Racial Problems*, edited by I. Rand Grigulevich and S. Y. Kozlob, 55–72. Moscow: Progress Pub.

Brown, P. L. 2011. "When the Uprooted Put Down Roots." *New York Times*, October 9. http://www.nytimes.com/2011/10/10/us/refugees-in-united-states-take-up-farming.html?_r=1&ref=patricialeighbrown.

Brownbridge, M. 2017. Vineland Research and Innovation Centre.

http://www.vinelandresearch.com/management-team/
michael-brownbridge-phd.

Bruno, D. 1994. "Build in Sustainable Development and They Will
Come: A Vegetable Field of Dreams." *Journal of Organizational
Change Management* 7, no. 4: 47–63.

Buchignani, N. 1977. "A Review of the Historical and Sociological
Literature on East Indians in Canada." *Canadian Ethnic Studies/
Etudes Ethniques au Canada; Calgary* 9, no. 1: 86.

———. 1985 *Continuous Journey: A Social History of South Asians in
Canada.* Toronto: McClelland and Stewart.

Burch, D., and G. Lawrence. 2009. "Towards a Third Food Regime:
Behind the Transformation." *Agricultural and Human Values* 26:
267–79. http://link.springer.com/article/10.1007/s10460-009-9219-4.

Buttel, F. H. 2001. "Some reflections on late twentieth century agrarian
political economy." *Sociologia Ruralis* 41, no. 2: 165–81.

Campigotto, R. M. 2010. "Farmers' Markets and Their Practices
Concerning Income, Privilege and Race: A Case Study of Wychwood
Artscape Barns in Toronto." MA thesis, University of Toronto.

Canadian Food Inspection Agency. 2009. "D- 99-06: Policy on the
Issuance of Phytosanitary Certificates." http://www.inspection
.gc.ca/plants/plant-protection/directives/exports/d-99-06/eng/
1323852257037/1323852328308.

Capella, L. M., and D. R. Arnold. 1993. "Acculturation, Ethnic
Consumers, and Food Consumption Patterns." *Journal of Food
Products Marketing* 1, no. 4: 61–79.

Caplan, P. 1997. "Approaches to the Study of Food, Health and
Identity." In *Food, Health and Identity*, edited by Pat Caplan, 1–31.
London: Routledge.

Carchedi, G. 1977. *On the Economic Identification of Social Class.*
London: Routledge and Kegan Paul.

Carter-Whitney, M. 2008. "Bringing Local Food Home: Legal,
Regulatory and Institutional Barriers to Local Food." *Friends of the
Greenbelt Occasional Paper Series.* Toronto.

Carter-Whitney, M., and S. Miller. 2010. *Nurturing Fruit and Vegetable
Processing in Ontario.* Toronto: Metcalf Foundation.

Chadeny, J. G. 1986. "India's Sikhs in Vancouver: Immigration, Occupation and Ethnic Adaptation." In *From India to Canada: A Brief History of Immigration: Problems of Discrimination, Admission and Assimilation*, edited by Chandrasekhar, 59–66. La Jolla, CA: Population Review.

Challinor, A. E. 2011. "Canada's Immigration Policy: A Focus on Human Capital." *Migration Immigration Source*. http://www.migrationpolicy.org/article/canadas-immigration-policy-focus-human-capital/.

Chan, A. 2011. *The Chinese in Toronto from 1878*. Toronto: Dundurn.

———. 2012. "From Chinatown to Ethnoburb: The Chinese in Toronto." Fifth International Conference of Institutes and Libraries for Chinese Overseas Studies (WCILCOS). https://open.library.ubc.ca/clRcle/collections/59585/items/1.0103074.

Chapman, G. E., S. Ristovski-Slijepcevic, and B. L. Beagan. 2011. "Meanings of Food, Eating and Health in Punjabi Families Living in Vancouver, Canada." *Health Education Journal* 70, no. 1: 102–12.

Charlebois, S. 2016. "High Food Prices Driving Some Shoppers Away from Fruits, Vegetables Study Says." http://www.cbc.ca/news/business/fruit-vegetable-prices-1.3617744.

Chia, A., and C. I. Costigan. 2006. "Understanding the multidimensionality of acculturation among Chinese Canadians." *Canadian Journal of Behavioural Science/Revue canadienne des sciences du comportement* 38, no. 4 (October): 311–24. http://dx.doi.org/10.1037/cjbs2006017.

Cho, L. 2010. *Eating Chinese: Culture on the Menu in Small Town Canada*. Toronto: University of Toronto Press.

Civitello, L. 2011. *Cuisine and Culture: A History of Food and People*. Hoboken, NJ: John Wiley and Sons.

Command Officers' Biographies. 2015. http://www.torontopolice.on.ca/bios/saunders.php.

Community Supported Agriculture. 2009. https://www.nal.usda.gov/afsic/community-supported-agriculture.

Cone, C. A., and A. Myhre. 2000. "Community-Supported Agriculture: A Sustainable Alternative to Industrial Agriculture?" *Human Organization* 59, no. 2: 187–97.

Conversi, D. 2010. "The Limits of Cultural Globalization?" *Journal of Critical Globalization Studies* no. 3: 36–59. http://criticalglobalization .com/Issue3/36_59_LIMITS_CULTURAL_GLOBALISATION_ JCGS3.pdf.

Cook, R., and U. Davis. 2000. "The Fresh Fruit and Vegetable Value Chain Faces New Forces for Change: AAEA Pre-Conference Workshop on Policy Issues and the Changing Structure of the Food System," Tampa Bay, Florida.

Cornforth, M. 1961. *Dialectical Materialism*. New York: International Publishers.

Crang, P., C. Dwyer, and P. Jackson. 2003. "Transnationalism and the Spaces of Commodity Culture." *Progress in Human Geography* 27, no. 4: 438–56.

Cuddleford, V. 2003. "When Organics Go Mainstream." *Alternatives Journal* 29, no. 4: 14–17.

Currie, E. 2016. Professor of business and economics, University of Guelph, discussed these details about okra production with Glen Filson, June 3.

Damböck, E. 2009. "Canada's New Global Past: Writing India's Past into Canadian History." In *Canadian Studies in Europe/Études canadiennes en Europe*. Vol. 8: *Dynamics of Canada: Studying Canada's Past and Current Realities. Dynamiques du Canada: Études sur les réalités contemporains et du passé au Canada*, edited by K. Battarbee and M. Buchart, 165–88. Brno, Czech Republic: Masaryk University, Faculty of Arts.

Darmon, N., and A. Drewnowski. 2008. "Does Social Class Predict Diet Quality?" *American Journal of Clinical Nutrition* 87, no. 5: 1107–17.

Darnhofer, I., R. Gretzmacher, and W. Schneeberger. 2006. "Modeling Farmers' Decisions: A Comparison Between HDM and CART for Oats-Vetch Adoption in the Ethiopian Highlands." *Die Bodenkultur* 48, no. 4: 271–80.

Das Gupta, T. 1994. "Political Economy of Gender, Race, and Class: Looking at South Asian Immigrant Women in Canada." *Canadian Ethnic Studies* 26, no. 1: 59–73.

Davidson, K. 2011. "Growers Bring Home Vegetables from the Global Village." *The Grower* 61, no. 5: 1–3.

Davis, K. 2015. "The Crossover Effects of Ethnocultural Vegetables: Examining the Guelph Consumers." MSc thesis, University of Guelph.

Daynard, K. 2016. "Locally Grown Okra Could Soon Be an Option for Ontario." *AgInnovation Ontario.* https://www.aginnovationontario .ca/en/locally-grown-okra-soon-option-canada/.

Delener, N. 1994. "Religious Contrasts in Consumer Decision Behaviour Patterns, Their Dimensions and Marketing Implications." *European Journal of Marketing* 28, no. 5: 36–53.

DeLind, L. B., and A. E. Ferguson. 1999. "Is This a Women's Movement? The Relationship of Gender to Community-Supported Agriculture in Michigan." *Human Organization* 58, no. 2: 190–200.

Desrochers, P., and H. Shimizu. 2012. *The Locavore's Dilemma: In Praise of the 10,000 Mile Diet.* New York: Public Affairs.

DeWeerdt, S. 2009. "Local Food: The Economics." *World Watch* 22, no. 4: 20–24.

Dindyal, S., and S. Dindyal. 2004. "How Personal Factors, Including Culture and Ethnicity, Affect the Choices and Selection of Food We Make." *Internet Journal of Third World Medicine* 1, no. 2. http:// www.ispub.com/journal/the-internet-journal-of-third-world -medicine/volume-1-number-2/how-personal-factors-including -culture-and-ethnicity-affect-the-choices-and-selection-of-food -we-make.html.

Dolan, C., and J. Humphrey. 2000. "Governance and Trade in Fresh Vegetables: The Impact of UK Supermarkets on the African Horticulture Industry." *Journal of Development Studies* 37, no. 2: 147–76.

Donald, B. 2009. "From Kraft to Craft: Innovation and Creativity in Ontario's Food Economy." *Martin Prosperity Institute Working Paper Series.* Toronto: Rotman School of Management, University of Toronto.

Donald, B., and A. Blay-Palmer. 2006. "The Urban Creative-Food Economy: Producing Food for the Urban Elite or Social Inclusion Opportunity?" *Environment and Planning* 38, no. 10: 1901–20.

Dunn, A. R., R. J. Sharkey, L. J. Manje, Y. Bouhlal, and M. R. Nayga Jr. 2011. "Socio-economic Status, Racial Composition and the Affordability of Fresh Fruits and Vegetables in Neighbourhoods of a Large Rural Region in Texas." *Nutritional Journal* 10, no. 6. http://www.ncbi.nlm.nih.gov/pmc/articles/PMC3033798/.

ECV Ontario. ECVOntario examines and prescribes policies for culturally appropriate foods in Ontario. evcontario2011.blogspot .com/.

Elford, E. 2013. "By the Numbers: Specialty Vegetable Markets in Ontario." http://onspecialtycrops.wordpress.com/2013/05/14/by-the-numbers-specialty-vegetable-markets-in-ontario/.

Fairholm, J., and L. Geggie. 1998. "Times They Are a-Changin': A New Wave of Youth Activism Promises a Broader Approach to Social Change. *Alternatives Journal* 24, no. 3: 10–17.

FAO. 2004. "Globalization of Food Systems in Developing Countries: Impact on Food Security and Nutrition." FAO food and nutrition paper. ftp://ftp.fao.org/docrep/fao/007/y5736e/y5736e00.pdf.

Feagan, R. B. 2007. "The Place of Food: Mapping Out the Local in Local Food Systems." *Progress in Human Geography* 31, no. 1: 23–42.

Feagan, R., and A. Henderson. 2009. "Devon Acres CSA: Local Struggles in a Global Food System." *Agriculture and Human Values* 26: 203–17.

Feagan, R. B., and D. Morris. 2009. "Consumer Quest for Embeddedness: A Case Study of the Brantford Farmers' Market." *International Journal of Consumer Studies* 33: 235–43.

Feagan, R. B., D. Morris, and K. Krug. 2004. "Niagara Region Farmers' Markets: Local Food Systems and Sustainability Considerations." *Local Environment* 9, no. 3: 235–54.

Fedoseyev, P. N., and V. Schneierson. 1977. *Leninism and the National Question.* Moscow: Progress Pub.

Ferris, L., and B. Behmann. 1994. "Farmers and Consumers Unite in Community Shared Agriculture." *Alternatives Journal* 20, no. 4: 9–10.

Fieldhouse, P. 1996. "Community Shared Agriculture." *Agriculture and Human Values* 13, no. 3: 43–47.

Filson, G. C. 1983. "Class and Ethnic Differences in Canadians' Attitudes Toward Native People's Rights and Immigration." *Canadian Review of Sociology and Anthropology* 20, no. 4: 454–82.

——. 2012. "Food Regimes, Sustainability and Ontario Agriculture: Issues and Overview." In *Agriculture and Environmental Security in Southern Ontario Watersheds,* edited by G. C. Filson, 1–16. New York: Nova Science Pub.

Filson, G. C., and B. Adekunle. 2011. "Towards a More Environmentally Secure Agriculture." In *Agriculture and Environmental Security in Southern Ontario Watersheds,* edited by G. C. Filson, 141–70. New York: Nova Science Pub.

Filson, G., B. Adekunle, S. Sethurathnam. 2011. *Growing International: Exploring the Demand for Culturally Appropriate Foods,* edited by S. Bloom and C. Young. Guelph: Farm Start.

Fine, B. 1994. "Towards a Political Economy of Food." *Review of International Political Economy* 1, no. 3: 519–45.

Fischler, C. 1988. "Food, Self, and Identity." *Social Science Information* 2: 275–92.

Food and Hunger Action Committee. 2000. *Planting the Seeds.* Phase 1 Report. Toronto: City of Toronto Food and Hunger Action Committee.

Foodland Ontario. 2013. *Fresh Ontario Food Is Closer Than You Think.* https://www.ontario.ca/foodland/foodland-ontario.

Foster, J. B., B. Clark, and R. York. 2010. *The Ecological Rift: Capitalism's War on the Earth.* New York: Monthly Review Press.

Friedland, W. H. 1994. "The Global Fresh Fruit and Vegetable System: An Industrial Organization Analysis." In *The Global Restructuring of Agro-Food Systems,* edited by P. McMichael, 173–89. Ithaca, NY: Cornell University Press.

Friedmann, H. 2005a. "Feeding the Empire: Pathologies of Globalized Agriculture." *Socialist Register* 41, no. 41: 124–43.

——. 2005b. "From Colonialism to Green Capitalism: Social Movements and Emergence of Food Regimes." In *New Directions in the Sociology of Global Development (Research in Rural Sociology and Development),* vol. 11, edited by Frederick H. Buttel and Philip McMichael, 227–64. Bingley: Emerald Group Publishing.

―――. 2007. "Scaling up: Bringing Public Institutions and Food Service Corporations into the Project for a Local, Sustainable Food System in Ontario." *Agriculture and Human Values* 24: 389–98.

Friedmann, H., and A. McNair. 2008. "Whose Rules Rule? Contested Projects to Certify "Local Production for Distant Consumers." *Journal of Agrarian Change* 8, no. 2–3: 408–34.

Friedmann, W. H., and P. McMichael. 1989. "Agriculture and the State System: The Rise and Decline of National Agriculture, 1870 to the Present." *Sociologia Ruralis* 29, no. 2: 93–117.

Gereffi, G., M. Korzeniewicz, and R. P. Korzeniewicz, eds.. 1994. *Commodity Chains and Global Capitalism*. Westport, CT: Greenwood Pub. Co.

Gibb, N. R. 2011. "Parallel Alternatives: Chinese-Canadians Farmers and the Metro Vancouver Local Food Movement." MA sociology thesis, Simon Fraser University.

Gibb, N., and H. Wittman. 2013. "Parallel Alternatives: Chinese-Canadian Farmers and the Metro Vancouver Local Food Movement." *Local Environment: The International Journal of Justice and Sustainability* 18, no. 1: 1–19.

Glanz, K., R. E. Patterson, A. R. Kristal, C. C. DiClemente, J. Heimendinger, L. Linnan, and D. C. Mclerran. 1995. "Stages of Change in Adopting Healthy Diets: Fats, Fibers and Correlates of Nutrient Intake." *Health Education* 22, no. 2: 261.

Glazer, N., and D. P. Moynihan, eds.1975. *Ethnicity: Theory and Experience*. Boston: Harvard University Press.

The Global City: Newcomer Health in Toronto. 2011. http://www1 .toronto.ca/wps/portal/contentonly?vgnextoid=227d5ce6dfb 31410VgnVCM10000071d60f89RCRD&vgnextchannel= 9553ebfc2bb31410VgnVCM10000071d60f89RCRD.

Goldman, A., R. Krider, and S. Ramaswami. 1999. "The Persistent Competitive Advantage of Traditional Food Retailers in Asia: Wet Markets' Continued Dominance in Hong Kong." *Journal of Macromarketing* 19, no. 2: 126–39.

Goodwin, H. L., S. W. Fuller, O. Capps, and O. W. Asgill. 1988. "Factors Affecting Fresh Potato Price in Selected Terminal Markets." *Western Journal of Agricultural Economics* 13, no. 2: 233–43.

Govindasamy, R. R., Van Vranken, and Sciarappa, W. 2007. *Demographics and the Marketing of Asian and Hispanic produce in the Eastern U.S.A.* New Brunswick, NJ: State University of New Jersey, Rutgers. Report # P-02903-2-07.

Granner, M. L., R. G. Sargent, K. S. Calderon, J. R. Hussey, A. E. Evans, and K. W. Watkins. 2004. "Factors of Fruit and Vegetable Intake by Race, Gender, and Age among Young Adolescents." *Journal of Nutrition Education and Behavior* 36, no. 4: 173–80.

Gray, M. 2016. "The Dark Side of Local." *Jacobin*, August 21. https:// www.jacobinmag.com/2016/08/farmGray, workers-local-locavore -agriculture-exploitation/.

Griffin, M. R., and E. A. Frongillo. 2003. "Experiences and Perspectives of Farmers from Upstate New York Farmers' Markets." *Agriculture and Human Values* 20, no. 2: 189–203.

Grisdale, C. 2016. "Foreign Workers Being Exploited to Grow Pot." *Toronto Star*, August 14, A11.

Gunst, J., G. Jaques, B. Jurjens, and T. McDowell. 2010. "The Inclusion of Ethnic Produce in Waterloo Region Food System." Paper prepared for Geography 429 Waterloo, University of Waterloo.

Guthman, J. 2008a. "Bringing Good Food to Others: Investigating the Subjects of Alternative Food Practice." *Cultural Geographies* 15: 431–47.

———. 2008b. "'If They Only Knew': Color Blindness and Universalism in California Alternative Food Institutions." *Professional Geographer* 60, no. 3: 387–97.

Hall, A., and V. Mogyorody. 2002. "The Marketing Practices of Ontario's Organic Farmers: Local or Global?" *Capitalism. Nature, Socialism* 13, no. 2: 3–34.

Hall, S. 1986. "Gramsci's Relevance for the Study of Race and Ethnicity." *Journal of Communication Inquiry* 10, no. 5: 4–27.

———. 1990. "Cultural Identity and Diaspora." In *Identity: Community, Culture, Difference*, edited by Jonathan Rutherford, 431–47. London: Lawrence & Wishart.

Hamilton, C., and K. Spence. 2008. "Ethnic & Specialty Food Exp. Key Industry Trends. Agri-Food Trade Service." Report by

Agriculture and Agri- Food Canada. http://www.ats.agr.gc.ca/eve/4568-eng.htm.

Hamlett, J., A. R. Bailey, A. Alexander, and G. Shaw. 2008. "Ethnicity and Consumption: South Asian Food Shopping Patterns in Britain, 1947–1975." *Journal of Consumer Culture* 8, no. 1: 91–115.

Han, T., and T. I. Wahl. 1998. "China's Rural Household Demand for Fruit and Vegetables." *Journal of Agricultural and Applied Economics* 30, no. 1: 141–50.

Haranandani, V. 2009. "Sustainable Agriculture in Canada and Cuba: A Comparison." *Environmental Development and Sustainability* 12: 763–75.

Haque, E. 2012. *Multiculturalism within a Bilingual Framework: Language, Race and Belonging in Canada.* Toronto: University of Toronto Press.

Harvey, C. 2016. "The Tangled Roots of Global Food Supply." *Toronto Star*, June 25, IN3.

Harvey, D. 1996. "Dialectics." In *Justice, Nature and the Geography of Difference*, 46–68. Oxford: Blackwell Pub.

Hatanaka, M., C. Bain, and L. Busch. 2005. "Third-Party Certification in the Global Agrifood System." *Food Policy* 30, no. 3: 354–69.

Health Canada, 2013. "Consumer Product Safety." http://www.hc-sc.gc.ca/cps-spc/pest/part/protect-proteger/publi-regist/index-eng.php.

Henry, F. 1998. "Two Studies of Racial Discrimination in Employment." In *Social Inequality in Canada*, 3rd ed., edited by J. Curtis, E. Grab, and N. Guppy, 226–35. Scarborough, ON: Prentice-Hall.

Herath, D., J. Cranfield, and S. Henson. 2008. "Who Consumes Functional Foods and Nutraceuticals in Canada? Results of Cluster Analysis of the 2006 Survey of *Canadians' Demand for Food Products Supporting Health and Wellness.*" *Appetite* 51, no. 2: 256–65.

Higgens, V., J. Dibden, and C. Cocklin. 2008. "Building Alternative Agri-food Networks: Certification, Embeddedness and Agri-environmental Governance." *Journal of Rural Studies* 24, no. 1: 15–27.

Hill, D. G., 1977. *Human Rights in Canada: A Focus on Racism.* Ottawa: Canadian Labour Congress.

Hinrichs, C. C. 2000. "Embeddedness and Local Food Systems: Notes on Two Types of Economic Market." *Journal of Rural Studies* 16, no. 3: 295–303.

———. 2003. "The Practice and Politics of Food System Localization." *Journal of Rural Studies* 19, no. 1: 33–45.

Hinrichs, C. C., and K. S. Kremr. 2002. "Social Inclusion in a Midwest Local Food System Project." *Journal of Poverty* 6, no. 1: 65–90.

Hui, A. 2016. "Chop Sue Nation." *Globe and Mail*, June 22, Section L.

Hunt, A. R. 2006. "Consumer Interactions and Influences on Farmers' Market Vendors." *Renewable Agriculture and Food Systems* 22, no. 1: 54–66.

Imhoff, D. 1996. "Community Supported Agriculture: Farming with a Face on It." In *The Case Against the Global Economy (and for a Turn Toward the Local)*, edited by Jerry Mander and Edward Goldsmith, 425–34. San Francisco: Sierra Club Books.

Intergovernmental Panel on Climate Change (IPCC). 2016. "Projections of Future Changes in Climate." https://www.ipcc .ch/publications_and_data/ar4/wg1/en/spmsspm-projections-of .html#table-spm-3.

Islam, S., M. Jalaluddin, and N. S. Hettriarachchy. 2010. "Functional Foods in Health and Disease: Bio-active Compounds of Bitter Melon Genotypes (*Momordica charantia L.*) in Relation to Their Physiological Functions." *Functional Foods in Heals and Disease* 1, no. 2: 61–74.

Jackson, R. J., R. Minjares, K. S. Naumoff, B. P. Shrimali, and L. K. Martin. 2009. "Agriculture Policy Is Health Policy." *Journal of Hunger & Environmental Nutrition* 4, no. 3: 393–408.

Jago, R., T. Baranowski, J. C. Baranowski, K. W. Cullen, and D. Thompson. 2007. "Distance to Food Stores & Adolescent Male Fruit and Vegetable Consumption: Mediation Effects." *International Journal of Behavioral Nutrition and Physical Activity* 4: 35. https:// ijbnpa.biomedcentral.com/articles/10.1186/1479-5868-4-35.

Jedwab, J. 2016. "How Angus Reid, CBC Got It Wrong about Multiculturalism." *Toronto Star*, October 13, A13.

Kajumba, C. 2012. "Ethno-Cultural Vegetable Retail Analysis: Pricing, Structure and Market Information." MSc thesis, University of Guelph.

Kane, D. J., and L. Lohr. 1997. *Maximizing Shareholder Retention in Southeastern CSAs.* Portland, OR: D. Kane.

Kerton, S., and A. J. Sinclair. 2010. "Buying Local Organic Food: A Pathway to Transformative Learning." *Agriculture and Human Values* 27, no. 4: 401. http://link.springer.com/article/10.1007/s10460-009-9233-6.

Kirwan, J. 2004. "Alternative Strategies in the UK Agro-Food System: Interrogating the Alterity of Farmers' Markets." *Sociologia Ruralis* 44, no. 4: 395–415.

Kneen, B. 1991. *From Land to Mouth: Understanding the Food System.* Toronto: NC Press.

Koc, M., and J. Welsh. 2002. "Food, Foodways and Immigrant Experience." In multiculturalism program, Department of Canadian Heritage at the Canadian Ethnic Studies Association Conference, Halifax, November. http://canada.metropolis.net/EVENTS/ethnocultural/publications/aliments_e.pdf.

Krotki, K. J. 1989. Review: *From India to Canada: A Brief History of Immigration; Problems of Discrimination; Admission and Assimilation* by S. Chandrasekhar. *The Pakistan Development Review* 28, no. 1 (Spring 1989): 57–64.

Krug, K. 2004. "Farm Women and Localized Alternatives to Globalized Agriculture." *Canadian Women Studies* 23, no. 1: 129–34.

Kymlicka, W. 1995. *Multicultural Citizenship: A Liberal Theory of Minority Rights.* Oxford: Clarendon Press.

Laclau, E. 1977. *Politics and Ideology in Marxist Theory: Capitalism, Fascism, Populism.* Verso: London.

Laraia, B. A., A. M. Siega-Riz, J. S. Kaufman, and S. J. Jones. 2004. "Proximity of Supermarkets Is Positively Associated with Diet Quality Index for Pregnancy." *Preventive Medicine* 39, no. 5: 869–75.

Larsen, K., and J. Gilliland. 2008. "Mapping the Evolution of 'Food Deserts' in a Canadian city, Supermarket Accessibility in London, Ontario, 1961–2005." *International Journal of Health Geography* 7, no. 16: 1–16.

Latham, J., and T. Moffat. 2007. "Determinants of Variation in Food Cost and Availability in Two Socioeconomically Contrasting Neighbourhoods of Hamilton, Ontario, Canada." *Health Place* 13, no. 1: 273–87.

Lee, L. E., O. Niode, A. H. Simonne, and C. M. Bruhn. 2012. "Consumer Perceptions of Food Safety in Asian and Mexican Restaurants." *Food Control* 26, no. 2: 531–38.

Lee, W., and D. K. Tse. 1994. "Changing Media Consumption in a New Home: Acculturation Patterns among Hong Kong Immigrants to Canada." *Journal of Advertising* 23, no. 1: 57–70.

Lister, N. 2007. "Placing Food." In *Food*, edited by J. Knechtel, 150–85. Alphabet City Series. Toronto: Ryerson University.

"Local Food." 2013. http://www.omafra.gov.on.ca/english/about/localfood.htm.

Lumpkin, J. R., B. A. Greenberg, and J. L. Goldstucker. 1985. "Marketplace Needs of the Elderly: Determinant Attributes and Store Choice." *Journal of Retailing* 61, no. 2: 75–105.

MacGregor, A. 2011. "Metro Partners with Adonis." *Gazette*, October 27. http://www.montrealgazette.com/business/Metro+partners+with+Adonis/5612796/story.html.

Maertens, M., and J. F. M. Swinnen. 2007. "Standards as Barriers and Catalysts for Trade, Growth and Poverty Reduction." *Journal of International Agriculture, Trade and Development* 4, no. 1: 47–61.

Marotte, B. 2011. "Metro Forges into Ethnic Food Market with Stake in Adonis." *Globe and Mail*, October 26.

Martinez, S., M. Hand, M. Da Pra, S. Pollack, K. Ralston, T. Smith, S. Vogel, S. Clark, L. Lohr, S. Low, and C. Newman. 2010. *Local Food Systems: Concepts, Impact, and Issues.* Washington: United States Department of Agriculture.

Martins, Y., M. L. Pelchat, P. Pliner. 1997. "'Try It; It's Good and It's Good for You': Effects of Taste and Nutrition Information on Willingness to Try Novel Foods." *Appetite* 28, no. 2: 89–102.

Marx, K. 1959. *Capital.* Vol. III: *The Process of Capitalist Production as a Whole.* Moscow: Progress Pub.

Marzall, K., G. C. Filson, and B. Adekunle. 2011. "Multifunctionality, Environmental Security and Strong Sustainability." In *Agriculture and Environmental Security in Southern Ontario Watersheds*, edited by G. C. Filson, 17–34. New York: Nova Science Pub.

Masoudi, G. 1993. "Kosher Food Regulation and the Religion Clauses of the First Amendment." *University of Chicago Law Review* 60, no. 2: 668–70.

McFarlane, T., and P. Pliner. 1997. "Increasing Willingness to Taste Novel Foods: Effects of Nutrition and Taste Information." *Appetite* 28, no. 3: 227–38.

McMichael, P. 1995. *Food and Agrarian Orders in the World Economy.* Westport, CT: Praeger.

McMichael, P. 2005. "Global Development and the Corporate Food Regime." In *New Directions in the Sociology of Global Development: Research in Rural Sociology and Development*, vol. 2, edited by F. Buttel and P. McMichael, 265–99. Bingley, UK: Emerald Group Publishing.

———. 2009. "A Food Regime Genealogy." *Journal of Peasant Studies* 36, no. 1: 139–69.

———. 2016. "Commentary: Food Regime for Thought." *Journal of Peasant Studies* 43, no. 3: 648–70.

Mintz, S. W. 2013. "Foreword: Food for Thought." In *The Globalisation of Chinese Food*, edited by D. Y. H. Wu and S. Cheung, xii–xx. New York: Routledge.

Morland, K., S. Wing, and R. A. Diez. 2002. "The Contextual Effect of the Local Food Environment on Residents' Diets: The Atherosclerosis Risk in Communities Study." *American Journal of Public Health* 92, no. 11: 1761–67.

Mount, P. 2012. "Growing Local Food: Scale and Local Food Systems Governance." *Agriculture and Human Values* 29, no. 1: 107–21.

Moverman, O., with S. McQueen. (dir.). 2013. *12 Years a Slave: The Extraordinary True Story of Solomon Northup.*

Muller, M., A. Tagtow, S. L. Roberts, and E. MacDougall. 2009. "Aligning Food Systems Policies to Advance Public Health." *Journal of Hunger and Environmental Nutrition* 4, no. 3: 225–40.

Naidoo, J. C. 2003. "South Asian Canadian Women: A Contemporary Portrait." *Psychology and Developing Societies* 15, no. 1: 51–67.

Nawaratne, Y. 2011. "Ethnocultural Vegetables in Ontario: Understanding the Value Chain." MSc thesis, University of Guelph.

Neff, R. A., A. M. Palmer, S. E. McKenzie, and R. S. Lawrence. 2009. "Food Systems and Public Health Disparities." *Journal of Hunger & Environmental Nutrition* 4, no. 3: 282–314.

Nestle, M., R. Wing, L. Birch, L. DiSogra, A. Drewnowski, S. Middleton, and C. Economos. 1998. "Behavioral and Social Influences on Food Choice." *Nutrition Reviews* 56, no. 5: 50–64.

Nicholas, B. 2012. "What Role Does the Ontario Food Terminal Play in the Ethnocultural Food Market?" Ethno-Cultural Vegetables Ontario: Challenges and Opportunities of the Ethno-Cultural Vegetable Market, at the John Eccles Centre, University of Guelph, February 17.

O'Doherty J. K., and L. Holm. 1999. "Preferences, Quantities and Concerns: Socio-cultural Perspectives on the Gendered Consumption of Foods." *European Journal of Clinical Nutrition* 53, no. 5: 351–59.

Ontario Fruit and Vegetable Growers' Association (OFVGA). 2011. "The Voice of Ontario's Fruit, Vegetable and Greenhouse Producers for Over 150 Years." http://www.ofvga.org/growers.php.

Ontario Human Rights Commission. 2016. http://www.ohrc.on.ca/en/ontario-human-rights-code.

Paliyath, G. 2011. *The Health Benefits of Eating Locally Grown Ethno-Cultural Vegetables.* ECV Ontario video, March 29. http://evcontario2011.blogspot.com/.

Pearson, T., J. Russell, M. J. Campbell, M. E. Barker. 2005. "Do 'Food Deserts' Influence Fruit and Vegetable Consumption?—A Cross-sectional Study." *Appetite* 45, no. 2: 195–97.

Pelchat, M. L., and P. Pliner. 1995. "'Try It. You'll Like It': Effects of Information on Willingness to Try Novel Foods." *Appetite* 24, no. 2: 153–65.

Pentland, H. C. 1981. *Labour and Capital in Canada: 1650–1860.* Toronto: James Lorimer and Co.

Peterson, A. R., K. R. Sharma, S. T. Nakamoto, and P. S. Leung. 1999. "Production Costs of Selected Vegetable Crops in Hawaii (Cabbage, Cucumber, Green Onion and Lettuce)." *Agribusiness*, Report no. 13, 1–8. http://hdl.handle.net/10125/3209.

Pliner, P., M. Pelchat, and M. Grabski. 1993. "Reduction of Neophobia in Humans by Exposure to Novel Foods." *Appetite* 20, no. 2: 111–23.

Pollan, M. 2013. *Cooked: A Natural History of Transformation.* New York: Penguin.

Porter, C. 2013. "Food Inspectors Put Short Ruler on Word 'Local.'" *Toronto Star*, April 25, A4.

Porter, M. 2002. "Enhancing the Microeconomic Foundations of Prosperity: The Current Competitiveness Index." http://www.isc.hbs.edu/Micro_9201.pdf.

Poulantzas, N. 1975. *Classes in Contemporary Capitalism.* London: New Left Books.

Powell, M. L., Z. Zhao, and Y. Wang. 2009. "Food Prices and Fruits and Vegetable Consumption Among Young American Adults." *Health and Place* 15, no. 4: 1064–70.

Puduri, V. S., and R. Govindasami. 2011. "Asian Consumers' Willingness to Buy Locally Grown Ethnic Produce: A Study from East-Coast United States." *Journal of Sustainable* Agriculture 35, no. 5: 511–21.

Raikes, P., and P. Gibbon. 2000. "'Globalization' and African Export Crop Agriculture." *Journal of Peasant Studies* 27, no. 2: 50–93.

Raikes, P., M. F. Jensen, and S. Ponte. 2000. "Global Commodity Chain Analysis and the French *filière* Approach: Comparison and Critique." *Economy and Society* 29, no. 3: 390–417.

Ramanujam, N., N. Caivano, and S. Abebe. 2015. "From Justiciability to Justice: Realizing the Human Right to Food." *McGill International Journal of Sustainable Development, Law and Policy* 11, no. 1: 1–38.

Raynborg, H. M., and J. F. Rubiano. 2001. "Farmers' Decision Making on Land Use—the Importance of Soil Conditions in the Case of Rio Cabuyal Watershed, Colombia." *Geografisk Tidsskrift-Danish Journal of Geography* 101, no. 1: 115–30.

Reardon, T., and J. F. M. Swinnen. 2004. "Agri-food Sector Liberalisation and the Rise of Supermarkets in Former State Controlled Economies: A Competitive Overview." *Development Policy Review* 22, no. 5: 515–23.

Reardon, T., C. P. Timmer, C. B. Barrett, and J. Berdegue. 2003. "The Rise of Supermarkets in Africa, Asia and Latin America." *American Journal of Agricultural Economics* 85, no. 1: 1140–46.

Regmi, A., M. S. Deepak, J. L. Seale Jr., and J. Bernstein. 2001. "Cross-country Analysis of Food Consumption Patterns." In *Changing Structure of Global Food Consumption and Trade*, edited by A. Regmi, 14–23. Washington: United States Department of Agriculture Economic Research Service.

Reynolds-Zayak, L. 2004. "Understanding Consumer Trends Can Present New Opportunities." Consumer trend report. Agri-Processing Branch. Business & Innovation Alberta Agriculture, Food and Rural Development. http://www1.agric.gov.ab.ca/$department/deptdocs.nsf/all/sis8735/$file/8735.pdf?OpenElement.

Riediger, N. D, and M. H. Moghadasian. 2008. "Patterns of Fruit and Vegetable Consumption and the Influence of Sex, Age and Socio-demographic Factors among Canadian Elderly." *Journal of American College of Nutrition* 27, no. 2: 306–13.

Roberts, R. A. G. 2002. *China to Chinatown: Chinese Food in the West*. London: Reaktion Books.

Roos, E., K. Talala, M. Laaksonen, S. Helakorpi, O. Rahkonen, A. Uutela, and R. Prättälä. 2008. "Trends of Socioeconomic Differences in Daily Vegetable Consumption, 1979–2002 Trends of Socioeconomic Differences." *European Journal of Clinical Nutrition* 62: 823–33.

Rose, D., and R. Richards. 2004. "Food Store Access and Household Fruit and Vegetable Use among Participants in the US Food Stamp Program." *Public Health Nutrition* 7, no. 8: 1081–88.

Rossman, P. 2016. "Unpacking CETA." http://column.global-labour-university.org/2016/09/unpacking-ceta.html.

Roy, P. E. 1972. "The Oriental 'Menace' in British Columbia." In *The Twenties in Western Canada: Papers of the Western Canadian Studies Conference*, 243–58. Ottawa: University of Ottawa Press.

Royce, A. P. 1982. *Ethnic Identity: Strategies of Diversity*. Bloomington: Indiana University Press.

Rozin, P. 1990. "Acquisition of Stable Food Preferences." *Nutrition Reviews* 48, no. 2: 106–13.

Rozin, P., and D. Schiller. 1980. "The Nature and Acquisition of a Preference for Chili Pepper by Humans." *Motive, Emotion* 4, no. 1: 77–101.

Rozin, P., and H. Tuorila. 1993. "Simultaneous and Temporal Contextual Influences on Food Acceptance." *Food Quality and Preference* 4, no. 1: 11–20.

Rozin, P., and T. A. Vollmecke. 1986. "Food Likes and Dislikes." *Annual Review of Nutrition* 6, no. 1: 433–56.

Ruel, M. T., N. Minot, and L. Smith. 2005. "Patterns and Determinants of Fruit and Vegetable Consumption in Sub-Saharan Africa: A Multicountry Comparison." http://searo.who.int /entity/ dietphysicalactivity/publications/f&v_africa_economics.pdf.

Sage, C. 2013. "The Interconnected Challenges for Food Security from a Food Regimes Perspective: Energy, Climate and Malconsumption." *Journal of Rural Studies* 29: 71–80.

Saunders, C., and P. Hayes. 2007. *Air Freight Transport of Fresh Fruit and Vegetables.* Research Report no. 299. Christchurch, NZ: Agribusiness and Economist Research Unit, Lincoln University.

Schafft, K. A., E. B. Jensen, and C. C. Hinrichs. 2009. "Food Deserts and Overweight Schoolchildren: Evidence from Pennsylvania." *Rural Sociology* 74, no. 2: 153–77.

Schnell, S. M. 2007. "Food with a Farmer's Face: Community-Supported Agriculture in the United States." *Geographical Review* 97, no. 4: 550–64.

Seccombe, W. 2007. "A Home-Grown Strategy for Ontario Agriculture: A New Deal for Farmers, a New Relationship with Consumers." Toronto Food Policy Council, September. http://www.toronto.ca/ health/tfpc/pdf/omafra_policy_shift.pdf.

Serecon Management Consulting. 2005. *Canadian Food Trends to 2020: A Long Range Consumer Outlook.* Prepared for agriculture and Agri-food Canada, Ottawa, Ontario. Prepared by Serecon Management Consulting Inc., Edmonton, Alberta. http://www4.agr.gc.ca/ resources/prod/doc/agr/pdf/ft-ta_eng.pdf.

Sharkey, J. R., and S. Horel. 2008. "Neighborhood Socioeconomic Deprivation and Minority Composition Are Associated with Better

Potential Spatial Access to the Ground-Truthed Food Environment in a Large Rural Area." *Journal of Nutrition* 138, no. 3: 620–27.

Shepherd, R. 1999. "Social Determinants of Food Choice." *Proceedings of the Nutrition Society* 58, no. 4: 807–12.

Shi, Y., C. Cheng, P. Lei, T. Wen, and C. Merrifield. 2011. "Safe Food, Green Food, Good Food: Chinese Community Supported Agriculture and the Rising Middle Class." *International Journal of Agricultural Sustainability* 9, no. 4: 551–58.

Shimakawa, T., P. Sorlie, M. A. Carpenter, B. Dennis, and G. S. Tell. 1994. "Dietary Intake Patterns and Socio-demographic Factors in the Atherosclerosis Risk in Community Study." *Preventive Medicine* 23, no. 6: 769–80.

Skinner, J. D., B. R. Carruth, W. Bounds, and P. J. Ziegler. 2002. "Children's Food Preferences: A Longitudinal Analysis." *Journal of the American Dietetic Association* 102, no. 11: 1638–47.

Skinner, J., B. R. Carruth, J. Moran, K. Houck, J. Schmidhammer, A. Reed, F. Coletta, R. Cotter, and D. Ott. 1998. "Toddlers' Food Preferences: Concordance with Family Members' Preferences." *Journal of Nutrition Education* 30, no. 1: 17–22.

Smith, A. D. 2000. *The Nation in History: Historiographical Debates about Ethnicity and Nationalism.* Hanover: University Press of New England.

Smith, A. D., and J. B. MacKinnon. 2007. *The 100 Mile Diet: A Year of Local Eating.* New York: Random House.

Sobal, J., and C. A. Bisogni. 2009. "Constructing Food Choice Decisions." *Annals of Behavioral Medicine* 38, no. 1: 37–46.

Sriskandarajah, D. 2002. "The Migration-Development Nexus: Sri Lanka Case Study." Paper prepared for the Centre for Development Research study "Migration-Development Links: Evidence and Policy Options," Magdalen College, Oxford, UK.

Stahlbrand, Lori. 2006. "Bringing Local Food to the City: The Local Flavour Plus Approach." *Local Food Plus*, November 23. http://www.canurb.com/media/Presentations/BS_306/LStahlbrand231106.pdf.

Statistics Canada. 2006a. *2006 Census Area Profiles.* http://www12
.statcan.gc.ca/census-recensement/2006/dp-pd/prof/rel/Rp-eng
.cfm?LANG=E&APATH=3&DETAIL=0&DIM=0&FL=A&FREE
=0&GC=0&GID=0&GK=0&GRP=1&PID=94533&PRID=0
&PTYPE=89103&S=0&SHOWALL=0&SUB=0&Temporal=2006
&THEME=81&VID=0&VNAMEE=&VNAMEF=.

————. 2006b. *Population by Selected Ethnic Origins, by Province and
Territory* (Ontario) (2006 Census). http://www.statcan.gc.ca/
tables-tableaux/sum-som/l01/cst01/demo26g-eng.htm.

————. 2006c. http://www.statcan.gc.ca/tables-tableaux/sum-som/
l01/cst01/agrc42g-eng.htm.

————. 2006d. *Population by Selected Ethnic Origins, by Province and
Territory (2006 Census) (Ontario).* http://www.statcan.gc.ca/
tables-tableaux/sum-som/l01/cst01/demo26g-eng.htm.

————. 2006e. *Population by Selected Ethnic Origins, by Province and
Territory (2006 Census) (British Columbia).* http://www.statcan
.gc.ca/tables-tableaux/sum-som/l01/cst01/demo26k-eng.htm.

————. 2007. *Immigration in Canada: A Portrait of the Foreign-
Born Population, 2006 Census.* Catalogue no. 91-557-XIE.
Ottawa: Ministry of Industry. http://www12.statcan.ca/census
-recensement/2006/as-sa/97-557/pdf/97-557-XIE2006001.pdf.

————. 2011. *Census Profile—Census Metropolitan Areas and Census
Agglomerations (CMAs/Cas).* Catalogue no. 98-316-XWE2011001
-201. http://www12.statcan.gc.ca/census-recensement/2011/
dp-pd/prof/details/download-telecharger/comprehensive/
comp-csv-tab-dwnld-tlchrgr.cfm?Lang=E.

————. 2013a. *2011 National Household Survey: Immigration, Place
of Birth, Citizenship, Ethnic Origin, Visible Minorities, Language and
Religion.* May 8. http://www.statcan.gc.ca/daily-quotidien/
130508/dq130508b-eng.pdf.

————. 2013b. *Population Estimate.* http://www.statcan.gc.ca/
start-debut-eng.html.

————. 2013c. *Visible Minority Population and Top Three Visible Minority
Groups, Selected Census Metropolitan Areas, Canada, 2011.* http://

www12.statcan.gc.ca/nhs-enm/2011/as-sa/99-010-x/2011001/tbl/
tbl2-eng.cfm.

———. 2017. *Population of Census Metropolitan Areas.* http://
www.statcan.gc.ca/tables-tableaux/sum-som/l01/cst01/
demo05a-eng.htm.

Steele, P., A. Dobson, H. Alexander, and A. Russell. 1991. "Who Eats
What? A Comparison of Dietary Patterns among Men and
Women in Different Occupational Groups." *Australian Journal
of Public Health* 15, no. 4: 286–94.

Stevens, N. 2009. "Equipping Consumers to Buy Canadian Food."
The CFFO Commentary, July 24. http://www.christianfarmers
.org/main_news_commentaries/2009commentaries/July_24
_Buying_Canadian.pdf.

Strauss, M. 2014. "Grocers Call for Code of Conduct Between
Retailers, Suppliers." *Globe and Mail,* March 8. http://www
.theglobeandmail.com/report-on-business/grocers-call-for-code
-of-conduct-between-retailers-suppliers/article17381094/%7b%
7burl%7d%7d/?reqid=%7b%7brequest_id%7d%7d.

"Stroll's 2013 Locavore Index Ranks States in Terms of Commitment
to Local Foods." http://www.strollingoftheheifers.com/locavore
-index-2013/.

Sumner, J., H. Mair, and E. Nelson. 2010. "Putting the Culture
Back into Agriculture: Civic Engagement, Community and the
Celebration of Local Food." *International Journal of Agricultural
Sustainability* 8, no. 1–2: 54–61.

Swinnen, J. F. M. 2005. "When the Market Comes to You—or Not:
The Dynamics of Vertical Co-ordination in Agro-Food Chains in
Europe and Central Asia." World Bank Report, Washington. http://
www.eastagri.org/meetings/docs/meeting12/VC%20version%20
%2024%20Feb%202005.pdf.

Swinnen, J. F. M., and M. Maertens. 2007. "Globalization, Privatization,
and Vertical Coordination in Food Value Chains in Developing and
Transition Countries." *Agricultural Economics* 37, no. 1: 89–102.

Szczepanski, K. 2015. "The Sri Lankan Civil War." http://asianhistory
.about.com/od/Sri_Lanka/fl/The-Sri-Lankan-Civil-War.htm.

Talbani, A., and P. Hasanali. 2000. "Adolescent Females Between Tradition and Modernity: Gender Role Socialization in South Asian Immigrant Culture." *Journal of Adolescence* 23: 615–27.

Thurlow, J., S. Benin, S. Diao, H. Kalinda, and T. Kalinda. 2008. *Agricultural Growth and Investment options for poverty reduction in Zambia*. IFPRI discussion paper 00791. Washington, DC: IFPRI.

Tilman, D., C. Balzer, J. Hill, and B. L. Befort. 2011. "Global Food Demand and the Sustainable Intensification of Agriculture. Proceedings of the National Academy of Sciences of the United State of America." *PNAS* 108, no. 50: 20260–64.

Toronto Food Business Incubator (TFBI). "City of Toronto Food Sector." http://www.tfbi.ca/01trfs.htm.

Toronto Food Policy Council (TFPC). 1996. "Food Retail Access and Food Security for Toronto's Low-Income Citizens." Toronto Food Policy Council Discussion Paper Series. Discussion Paper no. 7, December. http://www.toronto.ca/health/tfpc_food.pdf.

Trienekens, J. H. 2011. "Agricultural Value Chains in Developing Countries: A Framework for Analysis." *International Food and Agribusiness Management Review* 14, no. 2: 51–82.

Van der Ploeg, J. D. 2010. "The Food Crisis, Industrialized Farming and the Imperial Regime." *Journal of Agrarian Change* 10, no. 1: 98–106.

Van Halem, E. 2012. *Eat Local, Taste Global: A Guide to Growing and Cooking World Crops*. Toronto: J. Sutton Communications.

Van Huylenbroeck, G., and D. Damasco-Tagarino. 1998. "Analysing Crop Choice of Philippine Vegetable Farmers with Multicriteria Analysis." *Journal of Multi-Criteria Decision Analysis* 7: 160–68.

Vertovec, S. 1999. "Conceiving and Researching Transnationalism." *Ethnic and Racial Studies* 22, no. 2: 447–562.

Ward, P. 1978. *White Canada Forever: Popular Attitudes and Public Policy toward Orientals in British Columbia*. Montreal: McGill-Queen's University Press.

Wardle, J., M. L. Herrera, L. Cooke, and E. L. Gibson. 2003. "Modifying Children's Food Preferences: The Effects of Exposure and Reward on Acceptance of an Unfamiliar Vegetable." *European Journal of Clinical Nutrition* 57, no. 2: 341–48.

Waugh, V. F. 1928. "Quality Factors Influencing Vegetable Prices." *Journal of Farm Economics* 10, no. 2: 185–96.

WCM Consulting Inc. 2008. *SWOT Analysis of the Ethnic Food Sector in Ontario.* Sharon, ON: Prepared for Agriculture and Agri-Food Canada.

Weir, K. A. 2014. *From Jicama to Jackfruit: The Global Political Economy of Food or Stirring the Pot: A History of African Cuisine.* New York: Taylor and Francis.

Weis, T. 2007. *The Global Food Economy: The Battle for the Future of Farming.* London: Zed Books.

Williams, R. 1977. *Marxism and Literature.* London: Oxford University Press.

———. 1981. *Culture.* Glasgow: Fontana Paperbacks.

Wilson, J. S., and V. Abiola. 2003. *Standards & Global Trade: A Voice for AFRICA.* Washington, DC: Report for The World Bank.

Winson, A. 1993. *The Intimate Commodity: Food and the Development of the Agro-industrial Complex in Canada.* Toronto: Garamond.

———. 2004. "Bringing Political Economy into the Debate on the Obesity Epidemic." *Agriculture and Human Values* 21: 299–312.

———. 2013. *The Industrial Diet: The Degradation of Food and the Struggle for Healthy Eating.* Vancouver: UBC Press.

Wittman, H., A. A. Desmarais, and N. Wiebe. 2010. *Food Sovereignty: Reconnecting Food, Nature and Community.* Edited by H. Wittman. Oxford: Pambazuka.

Wolf, M. M., A. Spittler, and J. Ahern. 2005. "A Profile of Farmers' Market Consumers and the Perceived Advantages of Produce Sold at Farmers' Markets." *Journal of Food Distribution Research* 36, no. 1: 192–201.

Wong, A. 1977. "A Study of the Relationship Between the Proficiency in English and Cultural Background of Chinese Immigrant Students and Their Educational and Social Development." PhD diss., University of Toronto (Ontario Institute for Studies in Education), Toronto.

World Health Report. 2003. "Diet, Nutrition and the Prevention of Chronic Diseases." WHO Technical Report Series 916. Report of a

Joint WHO /FAO Expert Consultation. http://whqlibdoc.who.int/trs/who_trs_916.pdf.

Wormsbecker, L. C. 2007. "Moving Towards the Local: The Barriers and Opportunities for Localizing Food Systems in Canada." Master of Environmental Studies thesis, University of Waterloo.

Wright, E. O. 1978. *Class, Crisis and the State.* London: New Left Books.

———. 1985. *Classes.* London: New Left Books.

Wright, L. T., C. Nancarrow, and P. Kwok. 2001. "Food Taste Preference and Cultural Influences on Consumption." *British Food Journal* 103, no. 5: 348–57.

Wu, D. Y. H., and S. Cheung. 2013. *The Globalization of Chinese Food.* New York: Routledge.

Xuereb, M., and E. Desjardins. 2005. "Towards a Healthier Food Community System for Waterloo Region. An Interim Report." http://chd.region.waterloo.on.ca/web/health.nsf/c56e308f49bfeb7885256abc0071ec9a/54ED787F44ACA44C852571410056AEB0/$file/Food%20Systems_Report.pdf.

Yi, S., V. Kanetkar, and P. Brauer. 2015. "Assessment of Heterogeneity in Types of Vegetables Served by Main Household Preparers and Food Decision Influencers." *Public Health Nutrition* 18, no. 15: 2250–758. doi:10.1017/S1368980015001019.

Zafiriou, M. 2005. "Food Retailing in Canada: Trends, Dynamics and Consequences." Paper presented at the Pacific Economic Cooperation (PECC) Meetings Pacific Food Systems Outlook, Kunming, China.

Ziegler, R. G. 1989. "A Review of Epidemiologic Evidence That Carotenoids Reduce the Risk of Cancer." *Journal of Nutrition* 119, no. 1: 116–22.

Index